137
Advances in Polymer Science

Editorial Board:
A. Abe · A.-C. Albertson · H.-J. Cantow · K. Dušek
S. Edwards · H. Höcker · J. F. Joanny · H.-H. Kausch
T. Kobayashi · K.-S. Lee · J. E. McGrath
L. Monnerie · S. I. Stupp · U. W. Suter
E. L. Thomas · G. Wegner · R. J. Young

Springer
*Berlin
Heidelberg
New York
Barcelona
Budapest
Hong Kong
London
Milan
Paris
Santa Clara
Singapore
Tokyo*

Grafting/ Characterization Techniques/ Kinetic Modeling

With contributions by
H. Galina, Y. Ikada, K. Kato, R. Kitamaru,
J. Lechowicz, Y. Uyama, C. Wu

Springer

This series presents critical reviews of the present and future trends in polymer and biopolymer science including chemistry, physical chemistry, physics and materials science. It is addressed to all scientists at universities and in industry who wish to keep abreast of advances in the topics covered.

As a rule, contributions are specially commissioned. The editors and publishers will, however, always be pleased to receive suggestions and supplementary information. Papers are accepted for „Advances in Polymer Science" in English.

In references Advances in Polymer Science is abbreviated Adv. Polym. Sci. and is cited as a journal.

Springer WWW home page: http://www.springer.de

ISSN 0065-3195
ISBN 3-540-64016-9
Springer-Verlag Berlin Heidelberg New York

Library of Congress Catalog Card Number 61642

This work is subject to copyright. All rights are reserved, whether the whole or part of the material is concerned, specifically the rights of translation, reprinting, re-use of illustrations, recitation, broadcasting, reproduction on microfilms or in other ways, and storage in data banks. Duplication of this publication or parts thereof is only permitted under the provisions of the German Copyright Law of September 9, 1965, in its current version, and permission for use must always be obtained from Springer-Verlag. Violations are liable for prosecution under the German Copyright Law.

© Springer-Verlag Berlin Heidelberg 1998
Printed in Germany

The use of registered names, trademarks, etc. in this publication does not imply, even in the absence of a specific statement, that such names are exempt from the relevant protective laws and regulations and therefore free for general use.

Typesetting: Data conversion by MEDIO, Berlin
Cover: E. Kirchner, Heidelberg
SPIN: 10573665   02/3020 - 5 4 3 2 1 0 - Printed on acid-free paper

Editorial Board

Prof. Akihiro Abe
Department of Industrial Chemistry
Tokyo Institute of Polytechnics
1583 Iiyama, Atsugi-shi 243-02, Japan
E-mail: aabe@chem.t-kougei.ac.jp

Prof. Ann-Christine Albertson
Department of Polymer Technology
The Royal Institute of Technolgy
S-10044 Stockholm, Sweden
E-mail: aila@polymer.kth.se

Prof. Hans-Joachim Cantow
Freiburger Materialforschungszentrum
Stefan Meier-Str. 21
D-79104 Freiburg i. Br., FRG
E-mail: cantow@fmf.uni-freiburg.de

Prof. Karel Dušek
Institute of Macromolecular Chemistry, Czech
Academy of Sciences of the Czech Republic
Heyrovský Sq. 2
16206 Prague 6, Czech Republic
E-mail: office@imc.cas.cz

Prof. Sam Edwards
Department of Physics
Cavendish Laboratory
University of Cambridge
Madingley Road
Cambridge CB3 OHE, UK
E-mail: sfe11@phy.cam.ac.uk

Prof. Dr. Hartwig Höcker
Lehrstuhl für Textilchemie
und Makromolekulare Chemie
RWTH Aachen
Veltmanplatz 8
D-52062 Aachen, FRG
E-mail: 100732.1557@compuserve.com

Prof. J. F. Joanny
Institute Charles Sadron
6, rue Boussingault
F-67083 Strasbourg Cedex, France
E-mail: joanny@europe.u-strasbg.fr

Prof. Hans-Henning Kausch
Laboratoire de Polymères
École Polytechnique Fédérale
de Lausanne, MX-C Ecublens
CH-1015 Lausanne, Switzerland
E-mail: hans-henning.kausch@lp.dmx.epfl.ch

Prof. T. Kobayashi
Institute for Chemical Research
Kyoto University
Uji, Kyoto 611, Japan
E-mail: kobayash@eels.kuicr.kyoto-u.ac.jp

Prof. Kwang-Sup Lee
Department of Macromolecular Science
Hannam University
Teajon 300-791, Korea
E-mail: kslee@eve.hannam.ac.kr

Prof. J. E. McGrath
Polymer Materials and Interfaces Laboratories
Virginia Polytechnic and State University
2111 Hahn Hall
Blacksbourg
Virginia 24061-0344, USA
E-mail: jmcgrath@chemserver.chem.vt.edu

Prof. Lucien Monnerie
École Supérieure de Physique et de Chimie
Industrielles
Laboratoire de Physico-Chimie
Structurale et Macromoléculaire
10, rue Vauquelin
75231 Paris Cedex 05, France
E-mail: lucien.monnerie@espci.fr

Prof. Samuel I. Stupp
Department of Materials Science
and Engineering
University of Illinois at Urbana-Champaign
1304 West Green Street
Urbana, IL 61801, USA
E-mail: s-stupp@uiuc.edu

Prof. U. W. Suter
Department of Materials
Institute of Polymers
ETZ,CNB E92
CH-8092 Zürich, Switzerland
E-mail: suter@ifp.mat.ethz.ch

Prof. Edwin L. Thomas
Room 13-5094
Materials Science and Enginering
Massachusetts Institute of Technology
Cambridge, MA 02139, USA
E-mail. thomas@uzi.mit.edu

Prof. G. Wegner
Max-Planck-Institut für Polymerforschung
Ackermannweg 10
Postfach 3148
D-55128 Mainz, FRG
E-mail: wegner@mpip-mainz.mpg.de

Prof. R. J. Young
Manchester Materials Science Centre
University of Manchester and UMIST
Grosvenor Street
Manchester M1 7HS, UK
E-mail: r.young@fs2.mt.umist.ac.uk

Contents

Surface Modification of Polymers by Grafting
Y. Uyama, K. Kato, Y. Ikada 1

Phase Structure of Polyethylene and
Other Crystalline Polymers by Solid-State ^{13}C NMR
R. Kitamaru 41

Laser Light Scattering Characterization of Special
Intractable Macromolecules in Solution
C. Wu 103

Mean-Field Kinetic Modeling of Polymerization:
The Smoluchowski Coagulation Equation
H. Galina, J. Lechowicz 135

Author Index Volumes 101 – 137 173

Subject Index 183

Surface Modification of Polymers by Grafting

Yoshikimi Uyama[1], Koichi Kato[2], and Yoshito Ikada[1*]

[*] To whom correspondence should be addressed.
[1] Research Center for Biomedical Engineering, Kyoto University, 53 Kawahara-cho, Shogoin, Sakyo-ku, Kyoto 606, Japan. E-mail: yyikada@medeng.kyoto-u.ac.jp
[2] Department of Chemical Science and Engineering, Faculty of Engineering, Kobe University, 1-1 Rokkodai-cho, Nada-ku, Kobe 657, Japan

Recently a variety of technologies have been proposed for improving surface properties of polymers. Among them is surface grafting of polymers. Although this is a rather new technology, polymer surface grafting offers versatile means for providing existing polymers with new functionalities such as hydrophilicity, adhesion, biocompatibility, conductivity, anti-fogging, anti-fouling, and lubrication. This review article describes various methods of grafting and grafted surface characterizations. Medical and non-medical applications connected with this polymer surface grafting are also presented referring to recent publications.

Keywords: Surface modification, Graft polymerization, Graft chain, Adhesion, Surface characterization

1	Introduction	3
2	Creation of Grafted Surfaces	4
2.1	Polymer Coupling Reactions	4
2.2	Graft Polymerization	5
2.2.1	Direct Chemical Modification	6
2.2.2	Ozone	7
2.2.3	γ-Rays	7
2.2.4	Electron Beams	8
2.2.5	Glow Discharge	8
2.2.6	Corona Discharge	11
2.2.7	UV Irradiation	11
3	Characterization of Grafted Surfaces	14
3.1	Structure of Graft Polymer Chains	14
3.2	Thickness of Graft Chains	15
4	Applications	19
4.1	Non-Medical Applications of Grafted Surfaces	19
4.1.1	Adhesion	19
4.1.2	Adhesive Interaction in Aqueous Media	21
4.2	Medical Applications of Grafted Surfaces	22

4.2.1	None-Fouling Surfaces	24
4.2.2	Physiologically-Active Surfaces	28
4.2.3	Slippery Surfaces	31
4.2.4	Tissue-Adhesive Surfaces	32

References 35

List of Symbols and Abbreviations

AAm	acrylamide
ATR	attenuated total reflection
DMAA	N,N-dimethylacrylamide
DNA	deoxyribonucleic acid
EG	ethylene glycol
EVA	ethylene-vinyl alcohol copolymer
GMA	glycidyl methacrylate
HAP	hydroxyapatite
HDPE	high density polyethylene
HEMA	2-hydroxyethyl methacrylate
LDPE	low density polyethylene
MAA	methacrylic acid
MMA	methyl methacrylate
NaSS	Sodium styrenesulphonate
NIPAM	N-isopropylacrylamide
PAAc	poly(acrylic acid)
PAAm	polyacrylamide
PAN	polyacrylonitrile
PDMAA	poly(N,N-dimethylacrylamide)
PDMS	polydimethylsiloxane
PE	polyethylene
PEG	poly(ethylene glycol)
PEO	poly(ethylene oxide)
PET	poly(ethylene terephthalate)
PHEMA	poly(2-hydroxyethyl methacrylate)
PMMA	poly(methyl methacrylate)
PP	polypropylene
PTFE	polytetrafluoroethylene
PU	polyurethane
PVA	poly(vinyl alcohol)
PVAc	poly(vinyl acetate)
PVC	poly(vinyl chloride)
PVP	poly(N vinylpyrrolidone)
SLE	systemic lupus erythematosus
XPS	X-ray photoelectron spectroscopy

1
Introduction

Most of the untreated surfaces of polymers used in industry are not hydrophilic but hydrophobic. It is, therefore, difficult to bond these nonpolar polymer surfaces directly to other substances like adhesives, printing inks, and paints because they generally consist of polar compounds. On the other hand, polymer surfaces generally adsorb proteins when brought into direct contact with a biological system, resulting in cell attachment or platelet aggregation. The protein adsorption and attachment of biological components trigger a subsequent series of mostly adverse biological reactions toward the polymeric materials. Therefore, the technologies for surface modification of polymers or regulation of the polymer surface interaction with other substances have been of prime importance in polymer applications from the advent of polymer industries. Some of the technologies have been directed to introduction of new functionalities onto polymer surfaces. The new functionalities introduced include improved surface hydrophilicity, hydrophobicity, biocompatibility, conductivity, anti-fogging, anti-fouling, grazing, surface hardness, surface roughness, adhesion, lubrication, and antistatic property. Theoretically, there is a large difference in properties between the surface and the bulk of a material and only the outermost surface is enough to be taken into consideration when the surface properties are concerned. However, this is not the case for polymer surfaces, as the physical structure of the outermost polymer surface is generally not fixed but continuously changing with time due to the microscopic Brownian motion of polymer segments. The polymer surface generally has high segmental mobility even at room temperature, in contrast to the rigid surface of metals and ceramics. This propensity for motion of polymer segments suggests that a polymer surface cannot be described as a depthless, two-dimensional plane, but is in reality a region with some thickness. Therefore, we should look at the polymer surface in a direction not only parallel to the plane but also perpendicular to the surface plane in order to understand well the nature of polymer surface. The surface of copolymers and blended polymers is likely to be heterogeneous in the horizontal chemical composition, but these polymers have a distribution of chemical composition also in the vertical direction. This vertical distribution is called a depth profile.

Although a variety of technologies have been proposed for improving surface characteristics, surface modification of polymers by grafting is a rather new technology. It offers versatile means for incorporating new functionalities into existing polymers. However, in spite of the potentially wide applications of such surface grafting technology, this has been applied only to a few cases in industry, probably because the basic studies required for the applications are still in their infancies. Another reason may have been that such a grafted polymer surface is relatively expensive to produce and is used mostly in aqueous environments which are rarely encountered in the conventional industrial applications of polymers. Examples of application in such aqueous environments include marine science, biotechnology, and biomedical engineering. As these fields have become active only in recent years, it is not surprising that polymer modification by surface grafting has not yet become popular in industry. Recently, Hoffman summarized surface modification technologies by means of physical, chemical, mechanical, and biological methods [1].

Characterization of the structure of grafted surface is crucial, because a better understanding of the grafted surface will be a key factor for further developing this new technology. In this article, various methods of grafting and surface characterizations, and applications associated with grafting from recent literature publications will be reviewed.

2
Creation of Grafted Surfaces

There are in principle two methods for producing grafted surfaces, as schematically illustrated in Fig. 1: direct coupling reaction of existing polymer molecules to the surface and graft polymerization of monomers to the surface. Each has its own advantages and disadvantages, as demonstrated below.

2.1
Polymer Coupling Reactions

If the polymer surface to be modified possesses reactive groups capable of combining other components, such as water-soluble polymer molecules, surface modification can be readily conducted by chemical coupling reaction. Numerous synthetic reactions are available for this purpose. Bergbreiter [2] reviewed various

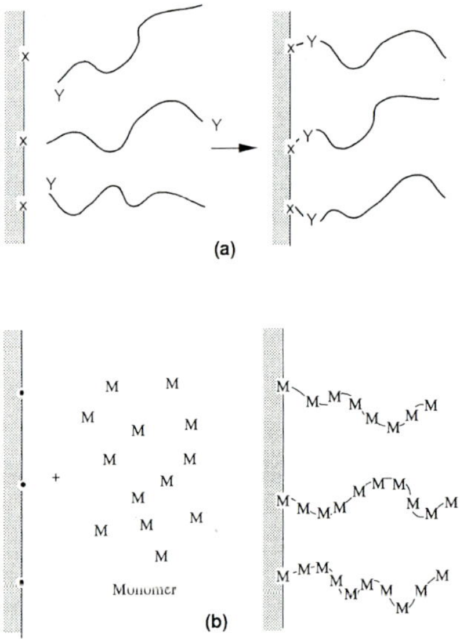

Fig. 1a,b. Creation of grafted surfaces by: **a** direct polymer coupling reaction; **b** graft polymerization

technologies and analytical methods associated with chemical modification of polymer surfaces. Kramer [3] proposed expressions for kinetics of grafting of a pure end-functional polymer melt to a reactive interface assuming that the free energy of reaction was very large and negative. Kishida et al. directly immobilized poly(ethylene glycol) (PEG) chains onto a cellulose surface through esterification [4]. Prior to coupling reaction, they derivatized the terminal hydroxyl group of PEG molecules to carboxylic acid using succinic anhydride. This PEG molecule with the terminal carboxylic acid was then chemically immobilized to the hydroxyl group on the cellulose surface using carbodiimide in non-aqueous media like toluene. Tezuka et al. [5] immobilized block copolymers onto poly-(vinyl alcohol) (PVA) and polyurethane (PU) surfaces. Polystyrene-polydimethylsiloxane (PS-PDMS) block copolymers containing vinylsilane or diol at the end of the chain ("macromer") were synthesized by living polymerization of styrene and dimethylsiloxane. Then the "macromer" was allowed to react chemically with the PVA or PU surface. Han et al. prepared PEO-grafted PU beads [6]. PU beads were first treated with hexamethylene diisocyanate in toluene with stannous octoate and subsequently chemically grafted with PEO in benzene in the presence of the same catalyst. Because a triethoxy silane group strongly bonds to a silanol group in the surface of silicon oxide, Clark et al. [7] converted a carboxy-terminated polystyrene to a triethoxy silane end polymer and carried out its surface grafting onto a cleaned silicone wafer. Ester exchange reaction was performed for grafting of 2-ethylhexyl acrylate and methyl methacrylate polymer [8] and the resulting copolymer was exchanged with PEG monomethyl ester.

2.2
Graft Polymerization

A variety of methods for graft polymerization onto different substrate surfaces will be described below from relatively recent reports. The methods include chemical graft polymerization and grafting with the use of high-energy radiation or oxidizing agents. For oxidizing a substrate surface, ozone, acidic treatments, and high-energy radiations in air can be used. Ionizing and nonionizing radiations as well as plasma treatment are usually selected for the high-energy sources. Strobel et al. [9] compared five gas-phase surface oxidation processes, i.e., corona discharge, flame, remote air plasma (Fig. 2), ozone, and combined UV/ozone treatment onto PP and poly(ethylene terephthalate) (PET) films. They revealed that flame, corona, and remote-plasma processes could readily oxidize the polymer surfaces, the O/C ratio being greater than 0.1 when treated for 0.04, 0.05–0.5, and 0.2 s, respectively. Of the five techniques, the frame treatment appeared to give the shallowest oxidation depth near the outermost surface region. In case that oxidized species is employed for initiation of graft polymerization, it should be noted that any method is not always applicable to all polymeric materials. For instance, fluorinated polymers such as polytetrafluoroethylene (PTFE) do not undergo significant oxidation by ozone and corona treatment, but active species for graft polymerization are effectively produced by glow discharge treatment of these polymers. Fluorinated polymers often suffer serious chain scission upon exposure to vacuum ultraviolet (VUV) radiation, but PET and polyethylene (PE)

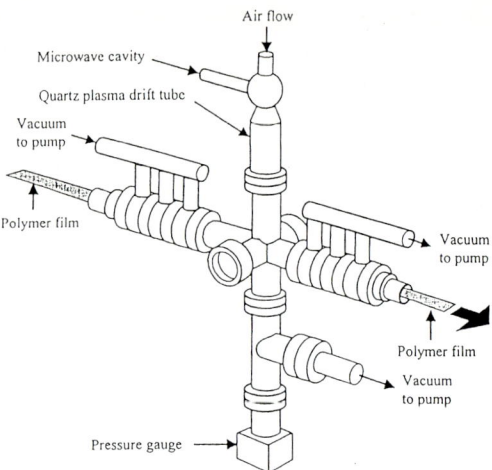

Fig. 2. Schematic diagram of the remote-plasma treater (Reproduced with permission from Strobel et al., J Adhesion Sci Technol 9: 365 Copyright (1995) VSP)

do not [10]. Ichijima et al. [11] compared the concentrations of peroxides introduced onto a poly(methyl methacrylate) (PMMA) substrate by different initiation methods, including ozone exposure, UV irradiation, corona discharge, and glow discharge. They evaluated their effect on the yield of graft polymerization of acrylamide (AAm) onto the oxidized PMMA surface. The most suitable initiation method for graft polymerization of AAm onto the PMMA substrate was found to be UV irradiation.

2.2.1
Direct Chemical Modification

Bergbreiter and Bandella [12] prepared a pH sensitive surface by immobilizing poly(acrylic acid) (PAAc) chains onto a PE surface. According to their technology, PE film surface was first functionalized by phenylpyrenylmethyl group, followed by graft polymerization of *t*-butyl acrylate monomer and subsequent hydrolysis to form PAAc. Pyrene groups were found to localize near the interfacial region, and the average degree of polymerization (DP) of acrylate polymer was estimated to be 20–30. They also reported an oxidation method by chromic sulfonic acid [13]. The esterified PE surface was then UV-irradiated in order to introduce radicals on the substrate, followed by graft polymerization of acrylonitrile.

The surface of polymeric materials such as polypropylene (PP), PS, polyacrylonitrile (PAN), and nylon was oxidized by immersing them in aqueous solution of oxidizing agents such as potassium peroxy disulfate under nitrogen purging at high temperatures [14]. Graft polymerization of water-soluble monomers such as AAm, methacrylic acid, and 3-aminopropyl methacrylate has been frequently performed in aqueous solution with the use of ceric ion, for instance, at

2×10^{-3} mol/l ceric ammonium nitrate in nitric acid at 50 °C. A dewaxed raw jute fiber was surface-grafted with methyl methacrylate (MMA) polymer using an initiator from the combination of IO_4^- and Cu^{2+} ions [15]. The initiation of grafting was attributed to radicals generated from cellulose through complexing between aqueous periodate ions and the anhydroglucose unit of cellulose molecule, followed by oxidation of the latter at the 1,2-glycol position. Grafting efficiency of 20–25% was obtained when an optimal polymerization condition ($[Cu^{2+}]$=0.001 mol/l and $[IO_4^-]$=0.005 mol/l) was used. A redox system consisting of $K_2S_2O_8$-$Na_2S_2O_3$-Cu^{2+} was used for surface graft polymerization of methacrylic acid (MAA) onto nylon 6 fibers and grafting took place mostly on the non-crystalline region of the substrate [16]. PET fibers were graft polymerized with a monomer mixture of AAm and MAA using AIBN [17]. The graft yield of the AAm polymer increased with the increasing fraction of MAA in the monomer mixture.

2.2.2
Ozone

PET, PE, PS, PP, and PC films were pretreated with ozone to introduce peroxy groups onto the substrate surface and AAm, AAc, and sodium styrenesulphonate (NaSS) monomers were graft polymerized by a near-UV induced technique [18]. The authors also performed surface graft polymerization of these monomers onto a polyaniline film. Although the pristine polyaniline film was effectively surface grafted by the near-UV induced method, combination of this method with ozone pretreatment enhanced the graft yield [19]. Buchenska [20] modified polyamide (PA6) knitted fabrics by graft polymerization of AAm and the hydroxy peroxide groups were introduced onto the fabrics surface by direct oxidation with oxygen in the temperature range from 120 to 160 °C prior to graft polymerization. The graft density obtained was 0.14×10^{-2} mol/g and subsequent treatment of the grafted fabrics with hydrazine or 3-bromopropionic acid improved both hydrophilic and antistatic properties of the fabrics.

2.2.3
γ-Rays

γ-Ray induced graft polymerization can be carried out by the following methods: 1. simultaneous irradiation and grafting through in situ formed free radicals, 2. grafting through peroxide groups introduced by pre-irradiation, and 3. grafting initiated by trapped radicals formed by pre-irradiation. Stannett [21] had reviewed the past 35 years' development of grafting via ionizing radiation. He recommended the use of electron beam radiation for polymer surface modification. Doué et al. [22] modified an industrial PP film by graft polymerization of AAc. The γ-ray irradiated PP film was immersed in a degassed aqueous solution of 50 wt% monomer and 0.1 wt% Mohr's salt at 85 °C for 2 h. Mohr's salt was added into the solution to prevent homopolymerization of the monomer and the peroxides yielded upon irradiation were quantitatively measured with the DPPH method. Poly(vinyl chloride) (PVC) powders were graft polymerized with AAc after the introduction of hydroperoxides onto the substrate surface by γ-ray irradiation in air at a dose rate of 84 MR/h from a 2100 Ci ^{60}Co source [23]. Percent graft yields were 100–200%, indicating that grafting was not localized in the surface region. The presence of ethanol, a chain transfer reagent, on the other hand, remarkably reduced the graft yield. Polymers of AAm and 2-hydroxyethyl methacrylate (HEMA) were grafted onto an ethylene-propylene copolymer rubber (EPR) by the radiation technique to improve the water uptake, wettability, and biocompatibility [24]. Fang et al. [25] conducted graft polymerization of vinyl acetate onto EPR by γ-irradiation to percentage graftings as high as 40%. Jan et al. [26] modified a porous poly(vinylidene fluoride) membrane to introduce positive charge by γ-irradiation in the presence of vinyl-triphenyl-phosphonium bromide monomer. In addition, PTFE-FEP copolymer [27], hollow fiber [28], PDM rubber [29], and PP fiber [30] were modified by graft polymerization with the radiation method.

2.2.4
Electron Beams

Polymers of MMA, AAc, and MAA were grafted onto an ultrahigh molecular weight polyethylene (UHMWPE) fiber surface after pretreatment with electron beam irradiation [31]. Sundell et al. [32] pretreated a PE film with electron beams to facilitate the graft polymerization of vinyl benzylchloride onto the substrate. The inner surface of porous PE hollow fiber had also been modified by grafting of glycidyl methacrylate (GMA) polymer after electron beam irradiation [33].

2.2.5
Glow Discharge

Low-temperature plasmas have been extensively applied to modify surface properties of polymer. Recently, surface modification by means of macromolecular plasma chemistry was reviewed by Denes [34], who also described the precise principle of plasma chemistry in detail. He pointed out that it is extremely difficult to understand and control the reaction mechanism of plasma because the relatively wide energy range (0.5–5 eV) of the glow discharge generates a very large number of charged and neutral molecular fragments. Various energy sources, such as UV, VUV, electron beams, and ozone are also involved in the plasma treatment if oxygen or air is present. Major applications of low-temperature plasma include plasma polymerization, which deposits a crosslinked thin polymeric layer on the substrate surface, and plasma treatment, which causes intensive oxidation in the surface region of the substrate.

It should be stressed that polymer surface can also be modified by graft polymerization utilizing free radicals or peroxides generated by the plasma treatment, similar to the effect of irradiation with high-energy radiations. The main difference between plasmas and high-energy radiations is the density of polymer radicals generated by exposure of a polymer substrate to these high-power sources. Sheu et al. [35] immobilized poly(ethylene oxide) (PEO) surfactants on hydropho-

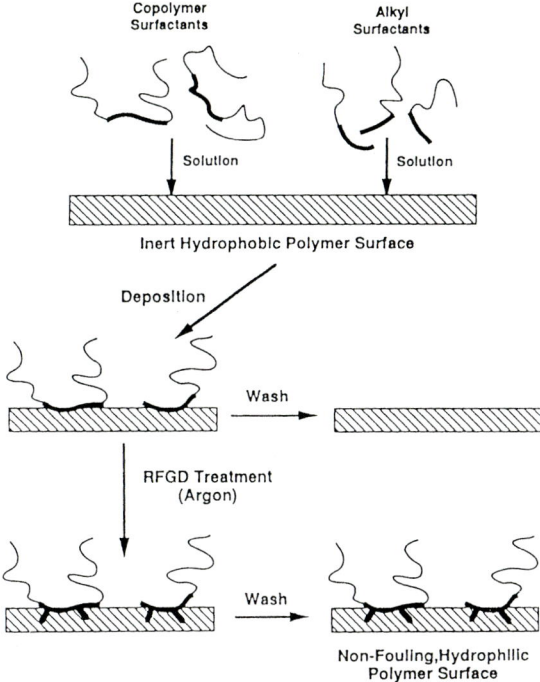

Fig. 3. Schematic diagram of argon glow discharge treatment for the immobilization of surfactants (Reproduced with permission from Sheu et al., Contact angle, wettability and adhesion, p 531 Copyright (1993) VSP)

bic polymer surfaces using the crosslinking by activated species of inert gases (CASING) technique. Low density polyethylene (LDPE) was first coated with alkyl PEO surfactants and then the substrate surface was crosslinked by Ar plasma treatment (see Fig. 3). Non-fouling (protein resistant) polymer surfaces were reported to be obtained by this method. A PEO-PPO-PEO triblock copolymer surfactant was also utilized for the CASING technique [35]. Hsie and Wu [36] exposed a PET film to the glow discharge of Ar plasma for 0.5–5 min at a gas pressure of 0.5 torr and a radio-frequency of 13.56 Hz. They then performed surface graft polymerization of AAc onto the plasma-treated film both in the vapor and liquid phases without exposing to air. Freshly synthesized PE powders were also treated with glow discharge, followed by surface graft polymerization of styrene without exposing to air [37]. ESR study revealed that graft polymerization started from the alkyl radicals introduced by the glow discharge and that more freshly synthesized powders were more amenable to form active radicals.

They also studied the melting characteristics of the powders with surface-grafted MMA polymer by DSC [38]. Grafting of Kevlar 49, or poly(*p*-phenylene terephthalamide), fiber with GMA and AAm polymers without exposing to air was studied by Yamada et al. [39]. According to their grafting system, Kevlar yarn was suspended in an upper plasma-irradiation vessel and irradiated for 1 min at an Ar gas pressure of 20 Pa, an rf frequency of 13.56 MHz, and an rf power of

100 W. The yarn was then immersed immediately in a GMA monomer solution placed at the lower side of the vessel (Fig. 4). They measured the adhesive strength between a resin matrix and the grafted Kevlar fiber. A similar method was employed by Yamaguchi et al. [40] for the graft polymerization of porous PE with methylacrylate monomer without contacting with air. It was found that radicals were formed in the porous substrate. Textile fabrics including cotton, cellulose acetate, and cupra ammonium cellulose were subjected to graft polymerization of HEMA, AAm, N-isopropylacrylamide (NIPAM), AAc, and 2-methoxyethyl acrylate (MEA) monomers in the absence of air after exposing them to 13.56 MHz rf Ar plasma [41]. In most cases, the graft yield obtained was 40–80%. Lai et al. [42] improved the adhesive strength of a silicone rubber against a conventional adhesive tape. The silicone rubber was pretreated with O_2 plasma and exposed to air in order to introduce peroxide groups onto the substrate surface. The pretreated rubber was immersed in aqueous monomer solution of AAm or AAc. There existed an optimum period of time for glow discharge treatment and graft polymerization with respect to both peroxide formation and adhesive force. Castner et al. [43] modified a silicone rubber by graft polymerization of AAc and characterized the surface changes using XPS, SIMS, and ATR-FTIR. Peroxides introduced onto the rubber surface were determined with the DPPH method. Lee and Shim [44] prepared a pH-sensitive poly(vinylidene fluoride) membrane by grafting with AAc polymer. The film was treated with Ar plasma for up to 3 min and then exposed to air, followed by graft polymerization of AAc in deaerated 20 wt% aqueous solution at 60 °C. The degree of polymerization (DP) for the homopolymer ranged from 32 to 94, depending on the plasma exposure time. A Kapton (poly(N,N'-oxydiphenylene pyromellitimide)) film pretreated with Ar plasma

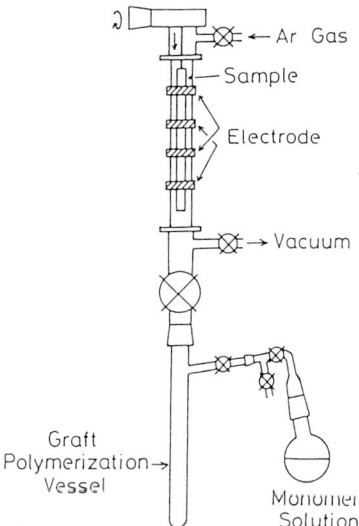

Fig. 4. Schematic representation of apparatus for plasma-graft polymerization (Reproduced with permission from Yamada et al., J Appl Polym Sci 60: 1847 Copyright (1996) John Wiley & Sons, Inc.)

was grafted with vinylimidazole polymer after exposing the plasma-treated film to air [45]. Grafting was also performed in benzene solution at 40–80 °C for 24 h to have a graft yield around 0.8 wt% in order to improve adhesion between the Kapton film and copper [46]. Qiu et al. [47] treated PU film with plasma at an rf of 2.45 GHz and a power of 300 W, exposed to air for 10 min, and then subjected the film to graft polymerization of PEG methacrylate (PEGMA) monomer. Hsiue et al. [48] pretreated a silicone rubber and a TPX film with Ar plasma and then subjected them to surface-grafting of polyHEMA to enhance cell adhesion onto the surfaces. These authors used also AAc [49], HEMA, and PEO for surface graft polymerization [50, 51]. Argon plasma treated PET, PE, and PP films were surface-grafted with N,N-dimethylacrylamide (DMAA) polymer to render the surface lubricious [52]. When the films were pretreated with benzoyl peroxide prior to graft polymerization, the graft amount was further increased. Plasma-induced graft polymerization was also performed on membranes of PAN and polysulphone using AAc as the monomer [53].

2.2.6
Corona Discharge

Both glow and corona discharges are currently used for surface modification of polymeric materials. Corona discharge treatment is by far simpler than glow discharge treatment, because the former technique does not require evacuation of the discharge system. However, corona treatment can result in more damage to the substrate polymers and the discharge conditions are more difficult to control than those of plasma treatment. The latter treatment is generally carried out under reduced gas pressure.

The surface of a low density PE sheet was modified by graft polymerization of ethylene glycol methacrylate with the use of corona discharge pretreatment [54]. A knife-type electrode, 1.5 mm away from the PE sheet surface, was connected to an rf generator of 50 W and 100 kHz. After corona treatment, grafting was performed at 80 °C for 4 h under continuous nitrogen bubbling to yield a comb-like PE surface. Various types of PP materials were subjected to surface treatment with corona discharge and the hydrophilicity and hydroperoxide yield were evaluated [55]. Isotactic PP was found to be more easily oxidized due to high crystallinity. The same authors also performed surface graft polymerization of AAm onto the pretreated PP surface. Lee et al. [56] grafted a corona-treated PE sheet with AAc polymer and subsequently subjected the surface to chemical reaction with various ionic reagents. Corona-treated PE films were grafted with various water-soluble polymers and the amounts of cell adhesion onto the grafted surfaces were compared [57].

2.2.7
UV Irradiation

The use of UV irradiation appears to be an excellent method for surface grafting of polymers because of its simplicity and cleanliness. Additional reasons for the suitability of the photochemical method for surface grafting of polymers are as follows: 1) photochemically produced triplet states of carbonyl compounds can abstract hydrogen atoms from almost all polymers so that graft polymerization may be initiated; 2) high concentrations of active species can be produced locally at the interface between the substrate polymer and the monomer solution containing a sensitizer when UV-irradiation is applied through the substrate polymer film; 3) in addition to the simplicity of the procedure, the cost of energy source is lower for UV radiation than for ionizing radiation.

Rånby et al. [58] suggested possible applications of surface modification to synthetic polymer fabrics by photo-induced graft polymerization. They performed graft polymerization of AAc and AAm onto the surface of fibers and films of PP and HDPE utilizing photo-initiators, such as benzophenone, 4-chlorobenzophenone, and hydroxycyclohexylacetophenone. Rånby [59] also discussed the adhesive properties of surface-modified materials. In order to functionalize the cellulose surface, graft polymerization of GMA was carried out in a Pyrex tube containing a cellulose sample, GMA monomer, and H_2O_2 [60, 61]. The H_2O_2 was decomposed by photo irradiation to produce hydroxyl radicals which might extract the hydrogen atom from polymer substrates to yield polymer radicals capable of initiating grafting. Graft yield depended on the species of photosensitizers used. They also performed graft polymerization using an initiator containing a cerium salt. A similar method was employed for graft polymerization of AAm on an ethylene-vinyl alcohol copolymer film [62] and for graft polymerization of AAc and 4-vinyl pyridine on a PE film [63]. In the latter case, the film was previously coated with benzophenone solution. Edge et al. [64] graft polymerized different monomers in the vapor phase onto a polyetherimide surface in the presence of a benzophenone solution in acetone. The identical method was also applied to the graft polymerization of some monomers onto low density PE and PET surfaces both in air and in N_2, and the surface properties were compared after characterizing with contact angle and XPS measurements [65]. PE, PP, PET, and PVA fabrics were surface-modified by photo-induced graft polymerization of AAc, AAm, glycidyl acrylate, and vinyl pyridine to improve the dyeing property of fabrics [66]. A block copolymer membrane consisting of PE, PS, and polybutadiene was subjected to graft polymerization of AAm in the presence of a small amount of isopropanol [67]. Water-soluble benzophenone derivatives were utilized to initiate UV-induced graft polymerization onto the membrane and the reaction mechanisms were discussed. Ulbrecht et al. [68] modified an ultrafiltration membrane with grafted PEG. Lee et al. [69] performed UV-induced graft polymerization of AAc, MAA, and NIPAM onto a polyamide membrane with combination of plasma treatment. Permeability of riboflavin depended on temperature when the NIPAM graft polymerized membrane was subjected to flow testing. A PE plate surface was surface-grafted with MAA polymer in the liquid phase after coating with a benzophenone (sensitizer) and poly(vinyl acetate) (PVAc) mixture [70,71]. After drying, the coated plate was immersed in an aqueous solution of MAA at 60 °C under a nitrogen atmosphere. Nakayama and Matsuda [72] performed surface photo-grafting using photo-reactive groups. Two different surfaces were grafted: a PS film whose surface was partially derivatized with *N,N*-diethyl-dithiocarbamyl groups prior to UV irradiation, and a PET film which was precoated with a photo-reactive copolymer cast from the toluene solution prior to UV exposure.

Photo-induced graft polymerization was also performed without using any photosensitizers and degassing procedure. Kang et al. [73] graft polymerized water-soluble monomers such as AAc, NaSS, and DMAA onto a PTFE film surface by near-UV irradiation with the combination of Ar plasma pretreatment. The plasma-pretreated film was exposed to air and then placed in an ampoule containing deaerated aqueous monomer solution, followed by exposure to a high-pressure mercury lamp. Grafting was also performed without the degassing procedure using riboflavin which consumed the dissolved oxygen in the course of photo irradiation. The surface of electroactive polymers such as polyaniline [74], polyimide [75], and polypyrrole film [76] was modified by similar graft polymerization of various water-soluble and ionic monomers. To improve the adhesion between epoxy molding compound (EMC) and ball grid array (BGA) substrate in the packaging of microelectronics systems, the surface of the epoxy-based BGA was graft polymerized with GMA [77], as depicted in Fig. 5. The authors also modified the surface of PE, PET, and PS films, and characterized the surface

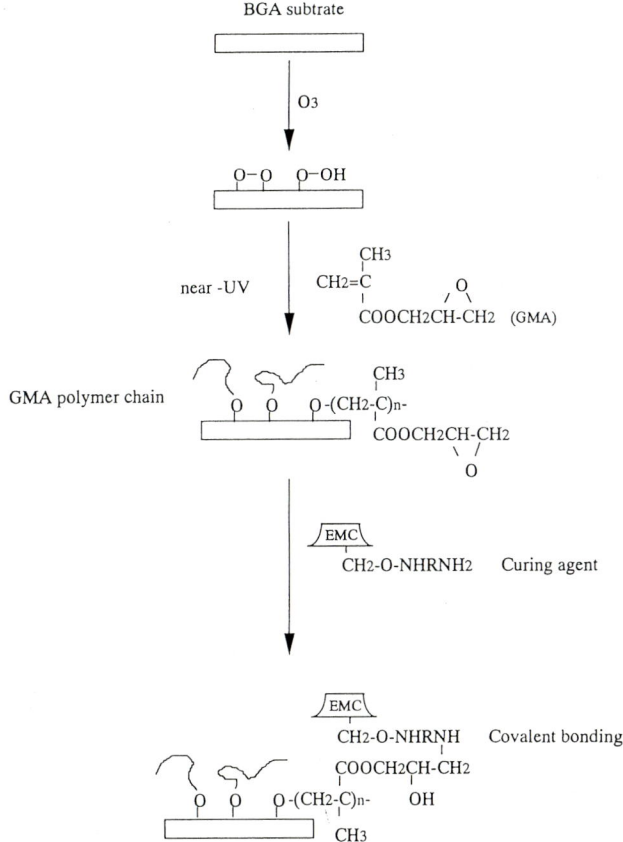

Fig. 5. Schematic diagram for the process of surface graft polymerization of GMA onto GA and subsequent reaction with amine curing agent in EMC

chemical structure [78]. Quaternary amines were introduced onto a PET film through photo-induced graft polymerization [79]. The UV-induced surface graft polymerization method was applied to UHMWPE and PET fibers for improved adhesion with an epoxy resin [80].

A laser beam was used for graft polymerization of AAc onto a tetrafluoroethylene-perfluoroalkyl vinyl ether copolymer film [81]. The film placed in contact with AAc solution was irradiated with KrF laser through the film to excite the film/solution interface. Surface composition of the grafted film determined by XPS revealed an extensive loss of fluorine atom and an increase of oxygen atom in addition to the presence of a C1s line shape, similar to that of AAc monomer. Mirzadeh et al. [82] used pulsed laser beam for the graft polymerization of AAm on a rubber surface in the presence of a photosensitizer, benzophenone, or AIBN.

Vacuum ultraviolet (VUV) irradiation was utilized for the graft polymerization of a mesogen-containing monomer onto fluorine-containing polymeric materials during the preparation of an anisotropic liquid crystal layer [83]. The grafted substrate comprised of two layers – a fluorine-containing polymer which provided good physical-mechanical properties, and a grafted chain layer which exhibited the usual features of nematic liquid crystal.

3
Characterization of Grafted Surfaces

Besides hydrophilicity and wettability, the atomic composition, overall molecular size, shape, mobility and conformational relaxation time of graft chains are of importance to the exploration physical and chemical structures of grafted surfaces. For these characterizations, a variety of analytical methods including spectroscopy, scattering, ellipsometry, and hydrodynamics have been proposed, in addition to computer simulation. Small angle neutron scattering is one of the most powerful techniques in determining the overall size of polymer chains and the spatial distribution of the chain segments [84]. The intensity and phase of scattered neutrons depend on the position and structure of the nucleus. On the other hand, neutron reflection, which is a relatively new technique, provides a unique characterization method for the atomic composition as a function of the depth of thin films. Fourier transform infrared spectroscopy coupled with attenuated total reflection (ATR-FTIR), Auger spectroscopy, and X-ray photoelectron spectroscopy (XPS) have become classical methods, but still provide very valuable information regarding the constituent elements and chemical structure near the surface region. The chemical derivatization techniques are also useful for surface analysis.

3.1
Structure of Graft Polymer Chains

Graft chains or long-chain polymer molecules attached by one end to a surface or the interface in contact with solvents are often called "polymeric brushes". A number of studies on the structure of such polymer chains have been undertaken using various theoretical means and computer simulations. The techniques employed include self-consistent field (SCF) theory [85–93], molecular dynamics (MD) [94, 95], and Monte Carlo (MC) simulation [96–100]. Detailed theoretical calculations revealed that long flexible polymer chains exhibit stretching under high graft densities, resulting in much greater height of graft chains than the radius of gyration. Irvine et al. [92] developed an SCF treatment of star-shaped polymer graft chains tethered to an impenetrable surface and used the model to calculate the near-surface equilibrium segment density profiles. At a moderate graft density, graft star molecules exhibited maximal stretching. Although these analyses promote an understanding of molecular structures under a variety of conditions or various environments, e.g., under θ point [100] and with or without solvent [95], it is not easy to prove experimentally the predicted molecular structure because of the difficulty in preparing the model graft chains. In addition, grafted surfaces are very difficult to characterize in detail because of the extremely low densities of graft chains on the surface. For instance, graft density amounts to only 0.1 μg/cm^2, if graft chains with a molecular weight of 1×10^5 are fixed as a monolayer at a frequency of one chain per 100 nm^2 on a surface. Conventional analytical methods are not sensitive enough to allow us to determine separately the length and the number density of graft chains fixed on a surface at such a low concentration, although several methods are available to evaluate the overall graft density which is a product of the chain molecular weight and the number density of the graft chains.

Concentration profiles of PS graft chains were studied by a neutron reflection method [101]. Graft chains consisted of end-functionalized deuterated polystyrene. Although the measurement was not performed under the true equilibrium conformation, the observed metastable state was in good agreement with that predicted from the SCF theory. The kinetics of the penetration of graft chains into the polymer matrix was also investigated.

Using angle-resolved XPS, the chemical composition and structure of film surfaces grafted with water-soluble polymers were investigated [78]. The XPS results showed that, in the case of substantially high grafting, the graft polymer chains penetrated or became partially submerged beneath a thin surface layer much richer in the substrate polymer. The surface structure was further confirmed using dynamic water contact angle measurements.

3.2
Thickness of Graft Chains

Some structural studies have been performed using simplified graft model chains. A long polymer chain occasionally possesses a number of functional sites which act as attachment points for the polymer chain to bind to a substrate surface as illustrated in Fig. 6. When a silica surface is placed in a melt or a solution of PDMS for a sufficiently long period of time, some segments become permanently bound to the solid surface. This type of graft layer is called pseudobrush or Guiselin brush. Cohen Addad et al. [102] determined the dry thickness of the pseudobrush prepared from various molecular weights of PDMS and concluded that the dry thick-

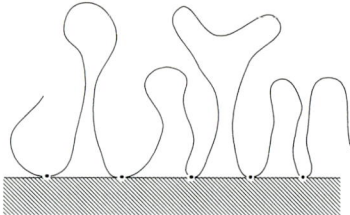

Fig. 6. Idealized representation of a "pseudobrush" or Guiselin brush

ness h had a linear correlation between $N^{1/2}$ and the monomer size a. h is proportional to the size of a mobile chain and N is a number of segments in one polymer chain. They determined the dry thickness by washing out all the unbound polymers by a good solvent followed by drying and weighing the residual polymer. Raudino and Zuccarello [103] derived a simple equation expressing the mean height of graft chains, having a statistical distribution of the internal conformations and hydrogen bonding, by combining statistical and quantum mechanical methods.

Wyart et al. [104] prepared a series of pseudobrushes by simultaneous adsorption of hydroxy-terminated PDMS onto a silicon wafer surface and reported that the dried thickness of these layers obeyed the scaling law of $h \approx N^{1/2} a \phi_0^{7/8}$, where ϕ_0 denotes the initial volume fraction of occupied polymer chains resulting from the interdigitation through a rubber matrix [105]. This relation was confirmed by other independent measurements such as neutron scattering [106], ellipsometry, and X-ray reflectance [107]. They also showed the existence of a maximum value for the 90° peel off adhesive energy at $\phi_0^{7/8}=0.4 \sim 0.5$. Cosgrove et al. [108] determined the hydrodynamic thickness by photon correlation spectroscopy. Silica particles having radii ranging from 0.08 to 0.2 µm were grafted with PS of Mw of 12 000–24 000. The graft density of PS was from 30 to 65 µg/cm². The hydrodynamic radius of particles was calculated from the Stokes-Einstein equation using the measured diffusion coefficient and the hydrodynamic thickness was obtained from the difference of the radii between the surface-grafted and the bare silica particles. The root mean square of the layer thickness could be fitted to the Alexander-de Gennes brush model, indicating that the chains were highly extended.

If a neutral layer of graft chains is present near a charged surface, the electroosmotic motion of the electrical double layer through the interface will be retarded. The reduction in electrophoretic mobility due to the neutral polymer chains assumes the hydrodynamic layer thickness. Janzen et al. [109] measured the electrophoretic mobility as a function of ionic strength for a PEG-grafted lipid bilayer. The mobility conformed to a model in which the PEG chains formed a surface layer of polymer with viscous drag arising from electroosmotic flow and provided the average layer thickness.

The hydrodynamic thickness of graft chains both in nonsolvated and solvated states was experimentally quantified by Webber et al. [110]. When a poly(2-

vinylpyridine)/polystyrene (PVP/PS) diblock copolymer was adsorbed in the pores of well-characterized mica membranes from toluene, the PVP block anchored on pore wall while the PS block imitated a terminally attached graft chain. They calculated the thickness of polymer chains from pressure drop as a function of flow rate through the membrane utilizing the modified Hagen-Poiseuille equation. Reversible extension of the polymer chains was observed when the solvent was altered from toluene to heptane or from heptane to toluene during the more than one month run. They further demonstrated that there was no shear thinning of the polymer layer over the shear rate range of 10^3–10^4 s^{-1}.

The flow rate retardation due to the attached graft chains was also observed using a porous membrane [111]. In this work, a porous poly(vinylidene fluoride) membrane was surface-grafted with PNIPAM which is soluble in water but has a lower critical solution temperature (LCST) around 31–33 °C. The flux of pure water through the grafted membrane varied by more than ten times between the temperatures above and below the LCST. The temperature sensitivity was reversible and reproducible. In this experiment, the thickness of graft chains could not be determined because no precise data for the pore size and density were available. The approximate layer thickness of water-swollen graft chains was directly measured using PNIPAM-grafted PU surface [112]. A virgin PU film became thicker under a temperature rise from 26 to 50 °C due to the thermal expansion, whereas the PNIPAM-grafted PU film became thinner under the same temperature rise because of the phase transition of the graft polymer chains. The thickness was of the order of a micrometer when estimated from the difference of the thickness change. The thickness of grafted DMAA and AAm polymer chains was estimated from the hydrodynamic flow using a PU tube (ID=0.65 mm, length=200 cm) [113]. The inner wall of PU tube was graft-polymerized with water-soluble monomers by an ozone pretreatment method. The flow rate of water through the PU tube in the range of laminar flow (Reynolds' number less than 10) was measured at a constant temperature. Under the simple assumption that the flow retardation was caused by the narrowing of the inner diameter of the tube in the presence of the surface graft layer, the graft thickness was estimated to be in the order of several micrometers. The flux change through a porous filter was also intensively discussed by de Gennes [114]. The increase in flux through a thin film composite membrane was reported by Mukherjee et al. [115] using a surface-modified membrane by chemical fluorination.

With respect to the dynamic structure of graft chains, direct and relevant experiment is difficult to perform, although some approaches have been proposed from theoretical analysis. Fytas et al. [116] reported on the dynamic behavior of brush-like chains using evanescent-wave dynamic light-scattering (EWDLS) which has an advantage in obtaining a time dependence of the deformation within a time scale of 10^{-7} to 10^3 s. They used an asymmetric PEO-PS block copolymer composed of two distinct blocks; a long PS block and a shorter PEO fragment. Dissolving the block copolymer in toluene, the PEO end-group was anchored onto a quartz prism surface, while PS chain imitated a stretched graft chain having the average layer thickness ranging from 45 to 130 nm. The evanescent wave prop-

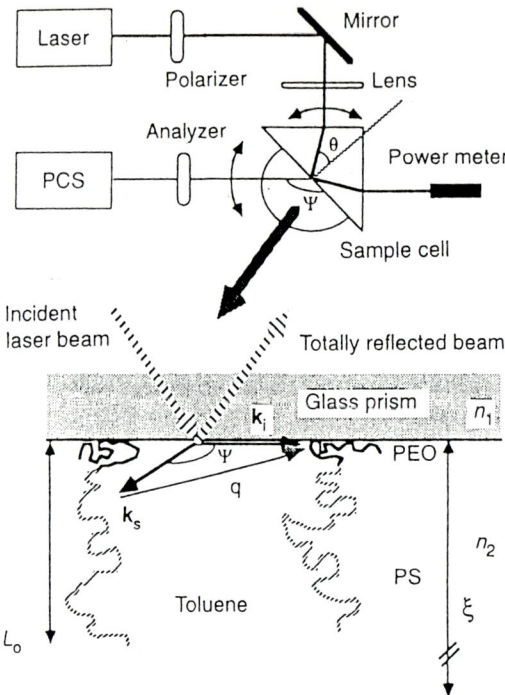

Fig. 7. Schematic diagram of the evanescent wave dynamic light-scattering instrumentation (Reproduced with permission from Fytas et al., Science 274: 2041 Copyright (1996) American Association for the Advancement of Science)

agated through the graft chains under the condition of total internal reflection (see Fig. 7).

Few studies were performed on the thickness of ionic graft chains. vonGoeler and Muthukumar [117] evaluated the height of a polyelectrolyte in solvents of varying salt concentrations in the high grafting regime. Integrating the electrostatic potentials over the directions parallel to the surface but neglecting the density fluctuation, they found that the dimensionless parameters $vN^{5/2}$ and $klN^{1/2}$ were valid for describing the long-ranged electrostatic interaction, where v, k, and l were the Coulombic interaction of strength, the inverse Debye screening length, and the Kuhn length, respectively. For instance, the chain height h was $l N^{1/2}$ for large values of both $vN^{5/2}$ and $klN^{1/2}$, and h was $l N^x$ with $1<x<3$ for large $vN^{5/2}$ if $vN^{5/2}/(klN^{1/2})^2$ was small. They also discussed the phase diagram of the scaling behavior and the transition of the polyelectrolyte in a poor solvent [118]. Experimental studies on the interaction between polyelectrolyte graft chains are also rare.

4
Applications

4.1
Non-Medical Application of Grafted Surfaces

4.1.1
Adhesion

When two chemically identical rubbers are brought into close contact with each other, a population of mobile chains at the rubber-rubber interface promote the adhesion between them. de Gennes [119] discussed the penetration of mobile graft chains in a rubber having an identical chemical structure. Such graft polymer chains are called "adhesion promoters" or "connectors". Raphaël and de Gennes [120] calculated the adhesion energy $G_{pull\ out}$ of the connectors grafted on one block and free at the other end, and thus being pulled out without any chemical rupture. They also compared the adhesion energy with that associated with chemical bond $G_{scission}$ and arrived at the following conclusion: $G_{scission}/G_{pull\ out}$ is nearly equal to U_x/U_v, where U_x is a typical van der Waals energy between two adjacent monomers and U_v represents the chemical bond energy. Some experimental results did not satisfy this relation, probably because the chains employed in the experiment were too short or because of the formation of bundles due to poor solvent.

Ultra-high modulus fibers such as aramid and carbon fibers have been currently utilized for composite material fabrication. Ultra-high modulus polyethylene (UHMPE) fiber is also applicable for composite fabrication because of the light weight in addition to its high modulus, vibration damping, and resistance to chemicals. However, this fiber has drawbacks such as poor interfacial adhesion with the polymer matrix of the composite because of highly hydrophobic nature of the fiber surface.

Li and Netravali [121] modified an UHSPE fiber surface using allylamine plasma deposition to improve its adhesion to epoxy resin. The shear strength of the modified surface increased by a factor of 2–3 when measured by pull-out test using a U-shaped epoxy resin. The epoxy resin and a curing agent were mixed and molded into U-shaped form without disturbing the fiber (see Fig. 8).

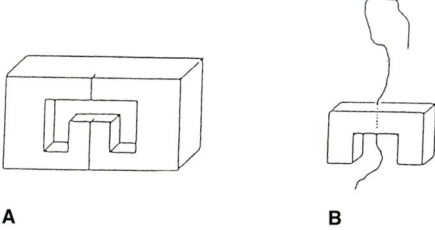

Fig. 8.a Silicone rubber mold. **b** Epoxy resin specimen with embedded fiber ready for single-fiber pull-out test (Reproduced with permission from Li and Netravali, J Appl Polym Sci 44: 333 Copyright (1992) John Wiley & Sons, Inc.)

The surface of Kevlar 49 fiber was chemically surface modified with epoxy groups containing reagents such as epoxy tetraglycidyl ether of diphenyldiamino methane [122]. It was found that the modified Kevlar fiber had enhanced adhesion to the epoxy matrix resin. Penn and Jutis [123] attached amine-terminated pendant groups to the surface of aramid fiber by chemical reaction. A single filament pull-out test used to assess the effect of the pendant groups on the fiber-epoxy matrix bond strength revealed that the presence of the pendant groups increased the adhesive performance. Rånby et al. have performed graft polymerization of AAc [124], glycidyl acrylate, and GMA [125] onto the surface of low density polyethylene with the use of UV irradiation in the vapor phase utilizing benzophenone as a photosensitizer. The surface of Kevlar fiber was also modified by Mori et al. [126] with the use of UV irradiation technique. To improve the Kevlar surface adhesion to epoxy resin, graft polymerization of AAm and GMA was performed onto the fiber surface. Following plasma treatment and the subsequent exposure to air, the fiber was placed in the monomer solution and subjected to UV irradiation. ATR-FTIR and XPS measurements revealed that the graft chains were present in the surface region of Kevlar fiber with the presence of epoxy groups on the surface of PGMA-grafted fiber. They also performed similar surface graft polymerization onto a UHMPE fiber pre-treated with Ar plasma to improve the wettability and adhesion [127]. A corona-treated PE sheet was surface-grafted with GMA and subjected to the measurement of adhesive shear strength against a cured epoxy resin [128]. The adhesive strength of the grafted PE was nearly twice as high as that of the corona-treated PE. They ascribed the enhanced adhesion between the surface-grafted PE and the epoxy resin to covalent bonding of the epoxy groups on the grafted surface to the amine moieties in the epoxy resin (see Fig. 9).

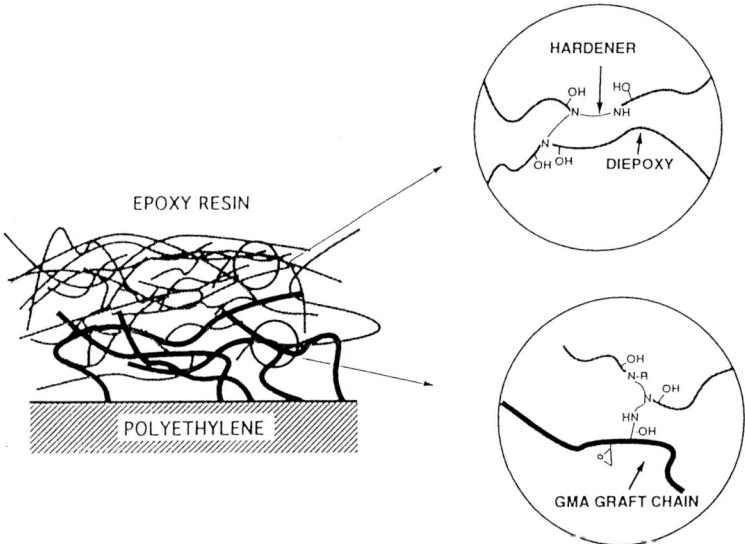

Fig. 9. Schematic representation of the structure of the interface between the GMA-graft polymerized PE and the cured epoxy resin

4.1.2
Adhesive Interaction in Aqueous Media

Polymer surfaces modified by graft polymerization of water-soluble monomers underwent substantial adhesion to another surface when they were brought into contact in the presence of water, under pressure and subsequently dried [129]. The surface having a larger graft density generally exhibited stronger adhesion when the water present at the interface dried up. Substantial interaction occurred almost instantaneously upon contact when one surface was grafted with an anionic polymer and the other grafted with a cationic polymer. The interaction between similarly charged surfaces was weak in the presence of water, probably because of the electrostatic repulsion operating between the similarly charged groups. However, even a similarly charged film pair exhibited adhesive force when water was completely dried up, probably because of entanglement of graft chains (see Fig. 10). They concluded that the adhesion strength depended on the surface graft density, the microstructure of the grafted surface, the nature of the interfacial interaction, and the adhesion (drying) time. Similar adhesion characteristics were observed by Kang et al. [130].

Wong et al. [131] measured directly the interaction potential between a tethered ligand and its receptor in aqueous media. Using a surface force apparatus, the interaction force-distance profile was determined between streptavidin immobilized on a lipid bilayer and biotin tethered to the distal end of lipid-PEG. Both lipid bilayers containing streptavidin and biotin were absorbed onto the surface of mica having a specific curvature. Both cationic and anionic polymer grafted

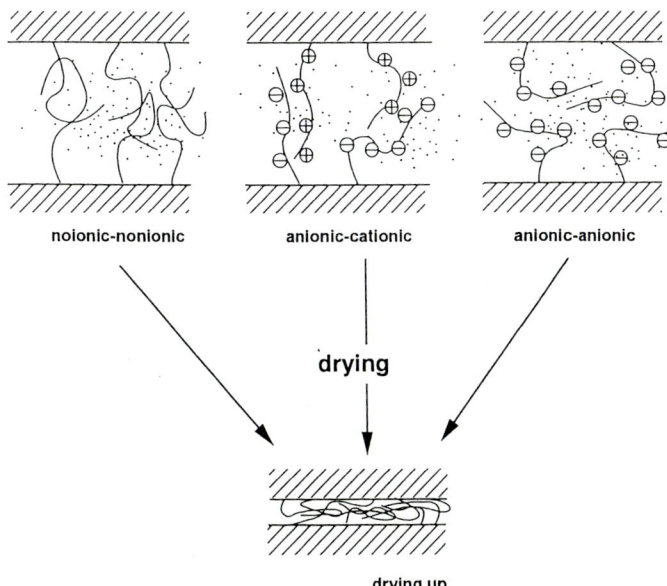

Fig. 10. Schematic representation of interaction between surfaces with graft chains

films exhibited a significant and instantaneous adhesive interaction in aqueous media towards solid surfaces, such as surface-modified cellulose film, soda glass, and untreated PET film, so long as the countersurface had oppositely charged zeta potentials [132]. It should be noted that none of surface pairs exhibited a significant attractive force, regardless of the sign of their zeta potentials, unless their surface was not graft polymerized with anionic or cationic polymers, as can be seen in Table 1. It may be postulated that the diffused polyelectrolyte graft chain layer can easily and almost directly bind to the oppositely charged surface region of the opposed substrate. In other words, the surface which showed a strong attractive force against a surface-grafted film may have a true surface potential the sign of which is opposite to that of graft chain. Detailed measurement of adhesion to a well-defined surface would make it possible to estimate the surface potential of solid materials not corresponding to the zeta potential.

More detailed electrostatic interaction in aqueous media between two surfaces grafted with ionic polymer was measured with the use of atomic force microscopy (AFM) [133]. The AFM tip surface was modified by graft polymerization of a cationic monomer after being coated with a cyanoacrylate polymer. A PET film which was the counter-surface of the tip for the AFM measurement was also modified by surface graft polymerization of ionic monomers. Appreciable adhesion was clearly observed in water between the grafted tip and the PET surface, while a repulsive interaction was noticed between the tip and the PET film when both were grafted with identically charged monomers. Addition of KCl to the medium exhibited a remarkable reduction in the adhesive interaction. These interactions were attributed to the Coulombic force between the grafted ionic polymer chains.

4.2
Medical Applications of Grafted Surfaces

Most biomedical materials are used in constant contact with living systems, such as blood, cells, and tissues. Since the material surface can undergo unfavorable biological responses when in contact with a recipient, most of the conventional surfaces need to be modified so that the materials can function as designed.

In the case of medical applications, some specific requisites usually dictate the modification technique. Among several methods investigated, grafting technique has advantages over others in several ways. They include easy and controllable introduction of new polymer chains with a high surface density and precise localization of the chains at the surface, while keeping the bulk properties unchanged. Furthermore, covalent attachment of polymer chains onto a polymer surface can avoid their delamination in aqueous media, and thus ensures long term stability of the introduced chains, in contrast to physically coated polymer chains. All these aspects give the rationale for applying the grafting process for surface modification. Accordingly, medical applications of graft polymers have attracted considerable attention in recent years.

In this section, we will highlight the use of the grafting technique for designing polymeric biomaterial surfaces that exhibit non-fouling property, selective protein adsorption, enhanced tissue adhesion, and minimum frictional damage to mucosa membranes.

Table 1. Attractive strength between cationic and ionic surfaces (N/cm^2)

	AAc-g-PET (-58)	Unmodified PET (-40)	Quartz glass (-33)	Cell-OCH$_2$COOH (-5.9)
Cell-DMAPAA (0)	1.8	<2.0 x 10^{-3}	6.3 x 10^{-3}	2.1 x 10^{-2}
DMAEMA-g-PET (+39)	27	4.3 x 10^{-3}	2.5 x 10^{-2}	5.8
Quartz glass (-33)	0	<2.0 x 10^{-3}	0	<2.0 x 10^{-3}

Note: the number in parentheses denotes the zeta potential in mV

4.2.1
Non-Fouling Surfaces

Prevention of protein adsorption is crucial to blood contacting devices including catheters, dialyzers, vascular grafts, blood containers, and oxygenators. When a material surface is brought into contact with blood, adsorption of serum proteins takes place as an initial event, which triggers successively the thrombogenesis and complement activation cascade via the classical pathway. Since these unfavorable foreign-body responses would lead to serious clinical problems, much effort has been devoted to the creation of non-fouling surfaces.

There is experimental evidence to support the versatility of surface grafting for minimizing protein adsorption as well as thrombogenesis. The grafting effect lasts for about a month or more, long enough for accomplishing the specific purpose of the devices, such as dialysis and oxygenation. However, the long-term durability of the grafting effect is still questionable [134, 135]. For example, vascular grafts and catheters for blood access require a non-fouling effect for a long duration of time if implanted permanently.

For creating non-fouling surfaces, various water-soluble polymers have been used for surface grafting. They include nonionic, hydrophilic polymers such as PAAm, PDMAA, PEG, EVA, and PHEMA. A polymer containing phosphorylcholine, a cell membrane constituent having zwitterions, was also used for the same purpose. Grafting of these polymers can be achieved by surface graft polymerization [136, 137], coupling reaction [4, 138–142], surface segregation [143, 144], and surface physical interpenetration [145]. Current advances in surface graft polymerization for non-fouling surfaces will be reviewed below. The details of other techniques are not within the scope of this section.

Fujimoto et al. demonstrated, through an ex vivo adsorption experiment using radio-labeled immunoglobulin G (IgG), that IgG adsorption onto a PU film was remarkably reduced by surface grafting of PAAm through glow discharge treatment [146] and ozone oxidation [147]. Interaction of the PU surface with platelets was greatly reduced by this modification, when assessed by an ex vivo arterio-venous (A-V) shunt experiment in rabbits [146, 147]. Ruckert and Geuskens [148] reported that surface graft polymerization of PVP effectively reduced adsorption of fibronectin onto a styrene-(ethylene-co-butene)-styrene triblock copolymer film. It was further reported that the surface of PU catheters with tethered PDMAA chains remained unfouled even after 3 weeks of implantation in the rabbit vena cava inferior [149]. Recently, Kishida et al. [150] used reverse transcription-polymerase chain reaction (RT-PCR) to examine the expression of interleukin-1β(IL-1β) mRNA secreted by macrophage-like cells (HL-60) cultured on a grafted PE surface as an index of inflammatory stimulation. They observed that cells cultured on the PAAm-grafted surface expressed a low level of IL-1β mRNA, indicating the non-fouling ability of the grafted surface.

Rejection of protein adsorption to the outermost grafted surface is attributed to a steric hindrance effect due to the tethered chains. A grafted surface in contact with an aqueous medium, a good solvent of the chains, has been identified to have a diffuse structure [57, 151, 152]. Reversible deformation of the tethered chains due to invasion of mobile protein molecules into the layer would lead to a repulsive force which is governed by the balance of entropic elasticity of the chains and osmotic pressure owing to the rise in the segment concentration. The overlapped repulsive force would prevent the direct contact of protein molecules with the substrate surface.

It is interesting to note that the extent of rejection of protein adsorption is related to the polymer graft density [146]. As shown in Fig. 11, the amount of adsorbed IgG increased gradually beyond the threshold (approximately 10 µg/cm^2), although the number of platelets adhered was kept at a low level. This result suggests that, when the graft layer is much larger in thickness than the protein radius and has a high water content, the protein molecules will readily migrate into the graft layer to be adsorbed, similar to protein molecules retained in the PAAm hydrogel in electrophoresis. Figure 12 illustrates schematically this protein sorption into a thick graft layer. Such sorption of plasma proteins into polymer matrices should be avoided, since the sorbed proteins would trigger the cascade of blood coagulation, similar to the plasma proteins in the stagnant, nonflowing blood.

Extensive grafting gives rise not only to an increase in the thickness of graft layer, but also to an increase in the volume fraction of graft segments in the layer. The latter also has a significant influence on protein adsorption. The water content of graft layer, which is a measure of volume fraction of graft segments in aqueous media, can be varied by changing the condition of surface graft polymerization. A well-defined model experiment using several hydrogels made of PVA, PAAm, PVP, PEG, and PHEMA [153] provided much information for understanding the interactions of grafted surfaces with blood components. The representative result obtained in this study is shown in Fig. 13. Apparently, all the hydrogels have an optimum water content at 80–95 wt% to minimize platelet adhesion, regardless of the chemistry of network chains. A graft layer with a water content higher than this threshold would not reject platelet adhesion. Kulik and Ikada

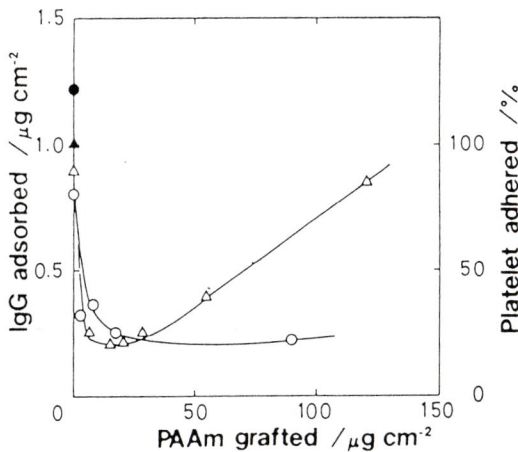

Fig. 11. Adsorption of ^{125}I-labeled IgG and adhesion of platelet to the surface of PAAm-grafted PU films as a function of graft density: (Δ) IgG (virgin, ▲), (○) platelet (virgin, •)

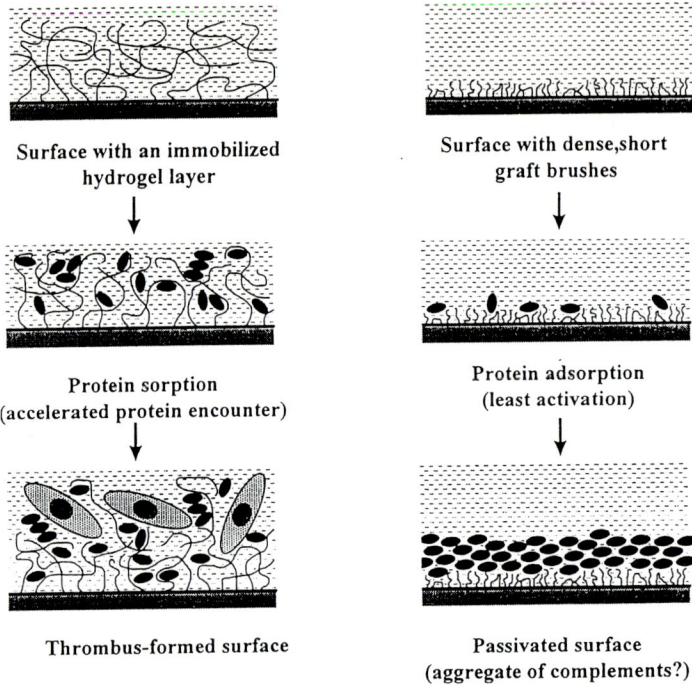

Fig. 12. Interaction of blood with surface having grafted water-soluble chains

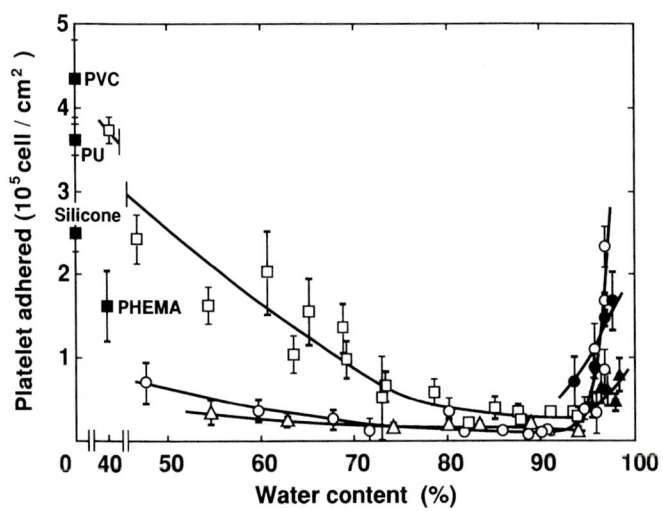

Fig. 13. Effect of water content of hydrogels on platelet adhesion: PVA (□), PAAm (○), methoxy-PEG methacrylate (Δ), PVP (●), and PEG (▲)

[153] attributed the phenomenon to the entrapment of platelets in the partially broken network structure of large pore size present at the surface of hydrogels. In contrast, when the water content of the hydrogels is below 80–95 wt%, the penetration of platelets through the hydrogel network seems to be greatly limited due to the high volume fraction of network chains. When the water content of the hydrogel becomes lower than 80 wt%, attraction of platelets by the outermost surface of the hydrogel will predominate probably due to the large free energy at the interface between the aqueous phase and the hydrogel surface. It was further demonstrated that carboxylic acids fixed to the hydrophilic polymer network of hydrogels served to reduce the number of platelets adhered. In contrast, cationic groups were found to have opposite effects on platelet adhesion [153]. This attractive interaction of cationic surface is operative not only for cell adhesion [56, 57] but also for protein adsorption [154]. These ionic effects may have relevance to ionic charges distributed over the surface of cells and protein molecules. As demonstrated by Kato et al. [154], protein adsorption onto polymer surfaces grafted with highly ionic polymer chains is primarily governed by the electrostatic attraction and repulsion between the charged surfaces and charged proteins. The ionically grafted polymer surface prevents adsorption of proteins with similar charge, but accelerates adsorption of proteins with opposite charge.

PEG molecules which are relatively nontoxic and capable of reducing the interactions between blood components and man-made materials, can also be tethered to a polymer surface through surface grafting. This has been achieved by free-radical polymerization of methacrylate monomers carrying a pendant PEG chain [47, 155–158]. The surface of a PU film was subjected to UV-induced graft polymerization of methoxy-PEG methacrylate monomers with 4, 9, and 23 units of ethylene glycol (EG) [156]. As shown in Fig. 14, the monomer with the shortest PEG length of only 4 EG units was most inert toward the blood components when the graft yield was largely reduced by the use of high concentrations

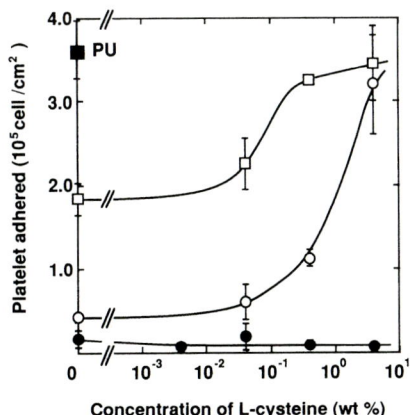

Fig. 14. Platelet adhesion to the PU films grafted with methoxy-PEG methacrylate at different concentrations of chain transfer agent. The number of EG units of the monomer: 4 (●), 9 (○), and 23 (□)

of a chain transfer agent. In addition, extraordinarily high graft yields were not effective in preventing protein adsorption, similar to the case of very low graft yields which were insufficient to reduce the interactions with proteins and platelets. Plasma-induced surface graft polymerization of PEG methacrylate was carried out onto bioresorbable poly(D,L-lactide) films [158]. Fibrinogen adsorption and platelet adhesion decreased upon grafting. Nevertheless, there was no significant effect on protein adsorption and platelet adhesion, so far as the number of EG units was maintained from 1 to 10.

In an attempt to synthesize hemocompatible dialysis membranes, a phosphorylcholine-containing methacrylate polymer was grafted to a cellulose membrane in a heterogeneous system using cerium ammonium nitrate [159, 160]. Kang et al. [161] and Qiu et al. [47] reported further immobilization of heparin, a naturally-occurring anticoagulant, onto surfaces grafted with acryloylbenzotriazole and ethylene glycol methacrylate polymers, respectively. It seems that the heparin-immobilized surfaces inhibit thrombus and fibrin formation by binding to the proteins involved in the clotting cascade.

As stated above, the non-fouling property is primarily based on the hydrophilicity of the graft chains. In this respect, multifunctionally hydroxylated PAAm, in which up to six hydroxyl groups were attached to each of the side chain terminal [162], and a glucose-containing methacrylate polymer, in which glucose residues were carried by each side chain through an ester bond [163], will be interesting candidates as these graft chains and capable of generating non-fouling surfaces.

4.2.2
Physiologically-Active Surfaces

Various reactive groups introduced onto polymer surfaces by direct grafting of functional polymers [73, 164, 165] or post-derivatization [166] of graft chains can be used to bind biologically active macromolecules, including proteins, polysaccharides, and deoxyribonucleic acids, onto the substrate surface. For example, primary amines, derivatized from amide groups of PAAm graft chains via Hofmann degradation as shown in Fig. 15, can be utilized for chemical immobi-

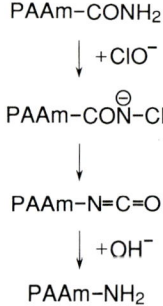

Fig. 15. Derivatization of the amide group of a PAAm graft chain into primary amine via Hofmann degradation

lization of proteins onto a substrate [166]. The amount of covalently immobilized proteins increased almost proportionally with the increase in graft yield, suggesting that protein molecules penetrated into and immobilized in the graft layer [164]. Such a surface possesses specific bioactivity associated with the immobilized biomacromolecules, allowing highly specific molecular recognition to be achieved. An example is selective adsorption of pathogenic proteins to the modified surface from blood. This selective adsorption has an important application as immunoadsorbents.

Immunoadsorption, an advanced therapeutic modality, focuses on detoxification of patient blood rich in high-molecular-weight pathogenic substances, mostly abnormal autoantibodies such as rheumatoid factors in rheumatoid arthritis (RA) and anti-DNA autoantibodies in systemic lupus erythematosus (SLE). Detoxification of these pathogens will be accomplished through extracorporeal perfusion of the patient plasma or whole blood over an affinity column made of immunoadsorbents. These adsorbents perform their function through the same mechanism as conventional affinity adsorption, where proteins in the liquid phase are adsorbed on the specific ligands immobilized onto an insoluble support.

A major problem associated with the current immunoadsorption is the low capacity of adsorbents, which can probably be attributed to the materials used as a solid support matrix. To solve this problem, an attempt was made to synthesize immunoadsorbents utilizing a solid support made of super fine PET microfibers [167, 168]. The use of such a fibrous support has great advantages over the conventional matrices, because this fiber is very large in specific surface area, excellent in mechanical strength, and biosafe.

The feasibility study of the immunoadsorbent synthesized through grafting of PAAc chains onto the PET fiber and the subsequent protein immobilization was conducted [167]. Three kinds of serologically active ligand proteins, including protein A, a specific antigen, and a specific antibody, were bound to the surface of PAAc-grafted PET fibers using 1-ethyl-3-(3-dimethylaminopropyl) carbodiimide (EDC). These adsorbents were tested for selective removal of IgG, specific polyclonal antibodies, and specific antigens. The binding capacity and specificity of the adsorbents were also evaluated in phosphate buffer solution and serum. The results are shown in Table 2, which provides useful information for applying immunoadsorbents to clinical systems.

An immunoadsorbent applicable to blood detoxification in patients with SLE has been synthesized by the grafting technology utilizing PET microfibers [168]. Double-stranded DNA was chosen as the ligand to be immobilized because the circulating pathogen in this case is polyclonal antibodies specific to DNA. The work was devoted to developing a method for DNA immobilization onto the grafted surface, because few publications have reported covalent immobilization of native DNA directly onto polymeric supports. Ample and stable immobilization of calf thymus double-stranded DNA, a highly anionic macromolecule, was carried out onto the PET fiber surface following the tethering of cationic graft chains with tertiary amines. A large capacity and a high specificity of anti-DNA antibody binding were found for the sera obtained from lupus mice and a human patient.

Table 2. The dissociation constant(K_d) and binding capacity(B_{max}) of adsorbents

Ligand	Protein A	Rabbit IgG	Anti-human IgG
Surface density of immobilized ligand(mol/cm^2) (μg/cm^2)	4.3 x 10^{-12} 0.18	1.6 x 10^{-11} 2.40	3.3 x 10^{-12} 0.59
From PBS			
K_d (10^{-7} mol/l)[a]	1.6≤2.1≤2.6	5.1≤20≤23	1.5≤2.5≤4.1
B_{max} (10^{-12} mol/cm^2)[a]	4.5≤5.1≤5.9	4.3≤13≤15	0.15≤0.19≤0.24
From serum			
K_d (10^{-7} mol/l)[a]	3.7≤6.6≤12	8.9≤12≤15	0.8≤6.5≤59
B_{max} (10^{-12} mol/cm^2)[a]	1.3≤1.8≤2.6	8.5≤11≤15	0.08≤0.14≤0.57

[a] Lower confidence limit ≤ mean ≤ upper confidence limit (p=0.05)

4.2.3
Slippery Surfaces

Polymer surfaces modified by immobilization of highly-hydrophilic graft chains exhibit lubrication in the hydrated state [169–173]. This property has spurred the effort to make the surface of tissue-contacting tubular devices such as catheters, cannulae, endoscopes, and cystoscopes lubricious. Most of the outer surfaces of body orifices, such as mouth, esophagus, nose, urethra, and vagina, into which tubular devices are often inserted, are covered with a mucosa layer. This layer consists mainly of hydrophilic glycoproteins producing highly viscous substances overlaying the endothelial cell basement. Lubricating the surface of the tubular devices will enable painless insertion, precise operation, and protection of the tissues from injuries.

Following ozone oxidation, the surface of a PU film was graft-polymerized with DMAA and the coefficient of kinetic friction (μ_k) for the fully hydrated, grafted films of two different graft densities was determined against a cleaned steel plate in distilled water as a function of the sliding velocity [174]. It was found that grafting of PDMAA effectively reduced the frictional force.

The same strategy was adopted to make the surface of a cystoscope lubricious [175]. It was covered with a PDMAA-grafted PU sheath of the same diameter as the cystoscope and the efficacy of the lubricated cystoscope was evaluated using rabbits by an in vivo test simulating cystoscope operation. As shown in Fig. 16, the maximal resistance force on the cystoscope was decreased by grafting. A histological study proved that urethral damage caused by rubbing against cystoscope was effectively reduced by this lubrication technique.

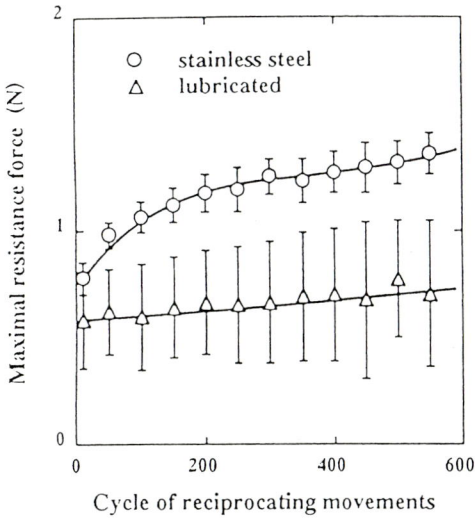

Fig. 16. Change in maximal resistance force for the cystoscope model during in vivo test. *Perpendicular bar* shows standard deviation

4.2.4
Tissue-Adhesive Surfaces

4.2.4.1
Soft Tissues

It is often demanded that the surface of polymeric biomaterials should exhibit permanent tenacious adhesion to soft connective and dermal tissues. However, conventional non-porous, polymeric materials will be encapsulated by a fibrous membrane generated de novo by surrounding fibroblasts, when subcutaneously implanted into the living body in contact with soft connective tissues. This is a typical foreign body reaction of the living system to isolate foreign materials from the host inside the body. On the other hand, it should be noted that the small gap present between a percutaneously-implanted device and the surrounding tissue provides a possible route for bacterial infection because of the lack of microscopic adhesion at the interface.

In contrast, when collagen, which is the most abundant extracellular component of connective tissues, is covalently immobilized onto an implant surface utilizing reactive groups introduced by surface graft polymerization and then implanted into the body, a fibrous membrane formed strongly adheres to the collagen-immobilized implant surface [176, 177]. Consequently, there remains no gap between the implant and the soft tissue. It was also reported that collagen immobilized onto a PE surface through PAAc graft chains had an inhibitory effect toward the tumor promoting activity [178].

Following the corona-discharge treatment of a silicone surface and the subsequent graft polymerization of AAc, type I atelocollagen was immobilized onto the grafted surface with the use of water-soluble carbodiimide [176, 177]. As depicted in Fig. 17, the immobilization reaction involves two steps, i.e., activation of carboxylic acids and the following nucleophilic substitution with prima-

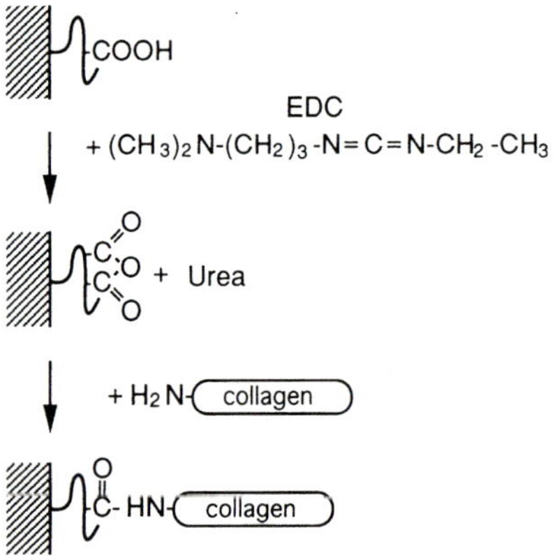

$$\text{—COOH}$$
$$\downarrow \quad \text{EDC}$$
$$+ (CH_3)_2 N\text{-}(CH_2)_3\text{-}N=C=N\text{-}CH_2\text{-}CH_3$$

$$\downarrow \quad + \text{ Urea}$$

$$\downarrow \quad + H_2 N\text{-}\boxed{\text{collagen}}$$

$$\overset{O}{\underset{\|}{-C}}\text{-HN-}\boxed{\text{collagen}}$$

Fig. 17. Collagen immobilization onto the PAAc-grafted polymer surface with the use of carbodiimide

ry amine moieties carried by collagen molecules. This method facilitated ample immobilization of collagen up to the order of 10 µg/cm² under mild conditions in aqueous media. This amount seems to be sufficient for the complete coverage of an implant surface with collagen. The feasibility study of this two-step method for covalent immobilization of collagen onto a PAAc-grafted surface was recently conducted by Lee et al. [179]. A simpler synthetic route for fabricating collagen-immobilized devices has been proposed recently [180]. The idea is based on the electrostatic attraction between different polyelectrolytes, through which collagen bearing a net positive charge forms a complex with an anionically grafted, polymeric surface. This method also provides stable immobilization of collagen.

The force required to pull out a collagen-immobilized silicone device was measured after three weeks of percutaneous implantation of this device into the rabbit back [177]. The results revealed that collagen immobilization remarkably enhanced the adhesion strength at the interface. Histological observation revealed that the immobilized collagen did not activate epidermal down-growth, but promoted the intimate adhesion at the material-tissue interface, although a relatively severe inflammatory reaction was observed in the initial stage of wound healing. As shown in Fig. 18, firm adhesion mediated by the immobilized collagen plays an important role as a barrier against infection.

The rationale for immobilizing collagen onto a biomaterial surface for its soft tissue adhesion is given by the fact that fibroblasts proliferate under either direct binding to an RGD (Arg-Gly-Asp) motif carried by the collagen molecule or indirect attachment mediated by tissue fibronectin bound to collagen in the living connective tissue. Plausibly, a collagen-immobilized polymer surface serves as a bed, similar to the natural extracellular matrix. Empirical confirmation of this mechanism was given by Tamada and Ikada [181]. The immobilized collagen was shown to promote the in vitro proliferation of fibroblasts and its metabolic activity in the initial stage of subculture.

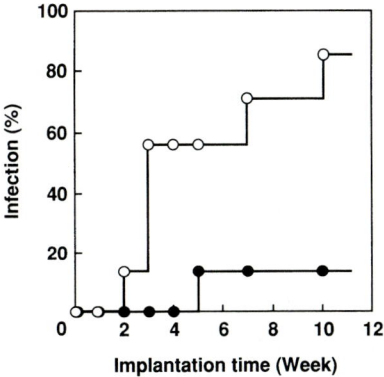

Fig. 18. Bacterial infection observed during percutaneous implantation of silicone devices into rabbits: (○) non-treated and (●) collagen-immobilized

Collagen immobilization onto a silicone rubber surface which underwent plasma-induced graft polymerization of PHEMA was also reported to have improved attachment and growth of corneal epithelial cells onto the rubber surface [48, 50, 182].

4.2.4.2
Hard Tissues

A flexible polymeric material having an ability to bond to hard tissues was synthesized through plasma-induced graft polymerization of a phosphate-containing monomer, methacryloyloxyethylene phosphate [183]. Potential applications of such a material are in artificial ligaments, tendons, intervertebral discs, and periodontia where strong adhesion to bones is needed. The dihydrogen phosphate of the monomer has an affinity for hydroxyapatite (HAP) which is the main inorganic component of bone tissues. When immersed in a simulated body solution of pH 7.4 containing calcium and phosphate ions at their supersaturated concentrations with respect to HAP, a polymer surface having the phosphate-containing chains promoted deposition of a thin HAP layer onto its surface. Figure 19 illustrates the schematic model proposed for deposition of an HAP layer. It was further demonstrated that bonding of the deposited HAP layer to the substrate surface was significantly improved by grafting of the phosphate-containing polymer chains. The study in simulated body environment led to the conclusion that the grafted surface has potentiality to bond firmly to bone tissues.

As in the case of strong fixation, rapid establishment of bone-implant adhesion is an important factor influencing the clinical consequences. It seems that this effect can be achieved by accelerating the HAP crystallization at the implant surface upon implantation. In this connection, an HAP layer was deposited in vitro onto a polymer surface grafted with the phosphate-containing polymer prior to implantation [184]. The pre-deposited HAP layer was fabricated by a solution-mediated wet method. The pre-deposited HAP served as a seed for HAP crystallites to grow rapidly, consuming calcium ions from the surrounding aqueous phase at a rate of 7 µg/cm^2/day, when brought into the physiological environment. Furthermore, enhanced calcium deposition at the HAP pre-deposited

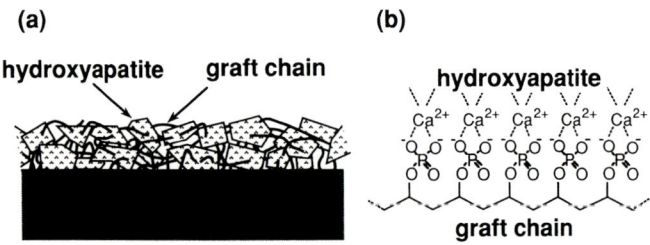

Fig. 19.a Proposed structure of the HAP firmly attached to the polymer surface having phosphate-containing graft chains. **b** Binding between the graft chains and the HAP crystal

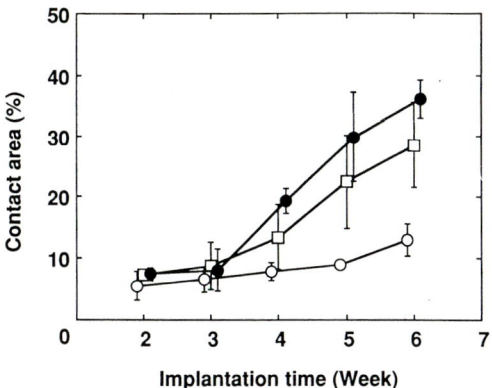

Fig. 20. Percentage of the PE rod surfaces in direct contact with the bone substrate established during the implantation into rat femora: (○) non-treated, (□) with phosphate-containing graft chains, and (●) with the HAP pre-deposited layer. *Errorbars* represent standard deviation

surface was also demonstrated in osteoblast cell culture [185]. The bone-bonding ability of polymer surfaces having phosphate-containing graft chains and a pre-deposited HAP layer was evaluated in vivo [186]. The surface-modified specimens were prepared using a PE rod and PET fabrics as substrates and implanted into the rat femur. As shown in Fig. 20, the results confirmed the bone-bonding ability of these surfaces and the accelerated HAP apposition by the pre-deposited HAP.

Acknowledgment. We are grateful for many fruitful discussions and valuable advice from Dr. Kang En-Tang of National University of Singapore.

References

1. Hoffman AS (1996) Macromol Symp 101: 443–454
2. Bergbreiter DE (1992) Chem Modif Surface 133–154
3. Kramer EJ (1995) Israel J Chem 35: 49–54
4. Kishida A, Mishima K, Corretge E, Konishi H, Ikada Y (1992) Biomaterials 17: 113–118
5. Tezuka Y, Nobe S, Shiomi T (1995) Macromolecules 28: 8251–8258
6. Han DK, Park KD, Ryu GH, Kim UY, Min BG, Kim YH (1996) J Biomed Mater Res 30: 23–30
7. Clark CJ, Jones RAL, Edwards JL, Shull KR, Penfold J (1995) Macromolecules 28: 2043
8. Zhu L, Gunnarsson O, Wesslen B (1995) J Polym Sci Part A, Polym Chem 33: 1257–1265
9. Strobel M, Walzak MJ, Hill JM, Lin A, Karbashewski E, Lyons CS (1995) J Adhesion Sci Technol 9: 365–383
10. Barbashev WA, Dorofeev YI (1992) Doklady Akademii Nauk SSSR 325: 730
11. Ichijima H, Okada T, Uyama Y, Ikada Y (1991) Makromol Chem 192: 1213–1221
12. Bergbreiter DE, Bandella A (1995) J Am Chem Soc 117: 10589–10590
13. Bergbreiter DE, Zhou J (1992) J Polym Sci, Part A, 30: 2049–2053
14. Bamford CH, Al-Lamee KG (1994) Polymer 35: 2844–2852
15. Ghosh P, Ganguly PK (1994) Polymer 35: 383–390

16. Rubtsov AE, Matveev UV, Chalykh A, Smirnova NV, Gabrielyan GA, Galbraikh LS (1991) J Appl Polym Sci 43: 729–736
17. Bergbreiter DE, Srinivas B, Xu G-F, Gray HN, Bandella A (1996) J Appl Polym Sci 59: 609–617
18. Loh FC, Tan KL, Kang ET, Neoh KG, Pun MY (1995) Eur Polym J 31: 481–488
19. Kang ET, Neoh KG, Tan KL, Ikada Y (1993) Synth Met 55: 1232–1237
20. Buchenska J (1995) J Appl Polym Sci 58: 1901–1911
21. Stannett VT (1990) Int J Radiat Appl Instrum, Part C 35: 82–87
22. Doué ILJ, Mermilliod N, Gandini A (1995) J Appl Polym Sci 56: 33–40
23. Pande CS, Single S, Gupta N (1996) J Appl Polym Sci 58: 1735–1739
24. Haddadi AV, Burford RP (1996) Radiat Phys Chem 47: 907–912
25. Fang Y -E, Lu XB, Wang SZ, Zhao Z, Fang F (1996) J Appl Polym Sci 62: 2209–2213
26. Jan D, Jeon JS, Raghavan S (1995) J Adhesion Sci Technol 8: 1157–1168
27. Bhattacharya A, De A, Bhattacharya SN (1994) Synth Met 65: 35–38
28. Mok S, Worsfold DJ, Fouda A, Matsuura T (1994) J Appl Polym Sci 51: 193–199
29. Katbab AA, Burford RP, Garnett JL (1992) Int J Raidat Appl Instrum, Part C 39: 293–302
30. Pilyugin V, Kritkaya DA, Ponomarev AN (1993) Vysokomol Soedin Ser A 35: 30–33
31. Zhang L, Liu Z, Yang N, Zhang A (1995) Zhongguo Fangzhi Daxue Xuebao 21: 88–93
32. Sundell MJ, Ekman KB, Svarfvar BL, Naesman NH (1995) Reac Polym 25: 1–16
33. Tsuneda S, Saito K, Furusaki S, Sugo T (1995) J Chromatogr A. 689: 211–218
34. Denes F (1997), Trend in Polym Sci 5: 23–31
35. Sheu MS, Hoffman AS, Feijen J (1993) In: Mittal KL (ed) Contact angle, wettability and adhesion. VSP Utrecht, pp.531–545
36. Hsie YL, Wu M (1991) J Appl Polym Sci 43: 2067
37. Tian J, Lin X, Huang B, Xu J (1995) Eur Polym J 31: 755–760
38. Tian J, Lin X, Huang B, Xu J (1995) J Appl Polym Sci 55: 741–746
39. Yamada K, Haraguchi T, Kajiyama T (1996) J Appl Polym Sci 60: 1847–1853
40. Yamaguchi T, Nakao S, Kimura S (1996) J Polym Sci, Polym Chem 34: 1203–1208
41. Zubaidi I, Hirotsu T (1996) J Appl Polym Sci 61: 1579–1584
42. Lai JY, Denq YL, Chen JK, Yuan LY, Lin YY, Shyu SS (1995) J Adhesion Sci Technol 9: 813–822
43. Castner DG, Mao G, Wang W, Grainger DW, McKeown P (1996) J Polym Sci, Polym Chem. 34: 141–148
44. Lee YM, Shim IK (1996) J Appl Polym Sci 61: 1245–1250
45. Inagaki N, Tasaka S, Matsumoto M (1995) J Appl Polym Sci 56: 135–145
46. Inagaki N, Tasaka S, Matsumoto M (1996) Macromolecules 29: 1642–1648
47. Qiu YX, Klee D, Plüster W, Severich B, Höcker H (1996) J Appl Polym Sci 61: 2373–2382
48. Hsiue GH, Lee S-D, Wang C-C, Shiue MH (1994) Biomaterials 15: 163–171
49. Lee SD, Hsiue GH, Kao CY (1996) J Polym Sci, Polym Chem 34: 141–148
50. Lee SD, Hsiue GH, Kao CY (1996) Biomaterials 17: 587–595
51. Hsiue GH, Lee SD, Chang PCT (1996) J Biomater Sci Polym Edn 7: 839–855
52. Onishi M, Shimura K (1993) J Photopolymer Sci Technol 6: 331–336
53. Ulbricht M, Belfort G (1996) J Membr Sci 111: 193–215
54. Jeong BJ, Lee JH, Lee HB (1996) J Colloid Interface Sci 178: 757–763
55. Novak I, Florian S (1995) J Mater Sci Lett 14: 1021–1022
56. Lee JH, Jung HW, Kang I-K, Lee HB (1994) Biomaterials 15: 705–711
57. Kishida A, Iwata H, Tamada Y, Ikada Y (1991) Biomaterials 12: 786–792
58. Rånby B, Gao FZ (1994) Polym Adv Technol 5: 829–836
59. Rånby B (1995) J Adhesion Sci Technol 9: 599–613
60. Kubota H, Ujita S (1995) J Appl Polym Sci 56: 25–31
61. Kubota H, Suzuki S (1995) Eur Polym J 31: 701
62. Kubota H, Sugiura A, Hata Y (1994) Polym Int 34: 313–317
63. Imaizumi M, Kubota H, Hata Y (1994) Eur Polym J 30: 979 983
64. Edge S, Feast WJ, Preston L, Walker S, Pacynko WF (1992) Polym Bull 27: 441–445
65. Edge S, Walker S, Feast WJ, Pacynko WF (1993) J Appl Polym Sci 47: 1075–1082
66. Feng Z, Icherenska M, Rånby B (1992) Angev Makromol Chem 199: 33–44
67. Ruckert D, Geuskens G (1996) Eur Polym J 32: 201–208

68. Ulbrecht M, Matuschewski H, Oechel A, Hicke HG (1996) J Membr Sci 115: 31–47
69. Lee YM, Ihm SY, Shim JK, Kim JH, Sung YK (1995) Polymer 36: 81–85
70. Yamada K, Kimura T, Tsutaya H, Hirata M (1992) J Appl Polym Sci 44: 993–1001
71. Bai GJ, Hy XZ, Yan Q (1996) Polym Bull 36: 503–510
72. Nakayama Y, Matsuda T (1996) Macromolecules 29: 8622–8630
73. Kang ET, Tan KL, Kato K, Uyama Y, Ikada Y (1996) Macromolecules 29: 6872–6879
74. Kang ET, Neoh KG, Tan KL, Uyama Y, Morikawa N, Ikada Y (1992) Macromolecules 25: 1959–1965
75. Loh FC, Lau CB, Tan KL, Kang ET (1995) J Appl Polym Sci 56: 1707–1713
76. Zhang X, Kang ET, Neoh KG, Tan KL, Kim DY, Kim CY (1996) J Appl Polym Sci 60: 625–636
77. Wang T, Kang ET, Neoh KG, Tan KL, Cui CQ, Chakravorty KK, Lim TB (1996) Mater Bull 31: 1361–1373
78. Loh FC, Tan KL, Kang ET, Uyama Y, Ikada Y (1995) Polymer 36: 21–27
79. Uchida E, Ikada Y (1996) J Appl Polym Sci 61: 1365–1373
80. Yuan M, Wu C, Shi F (1995) Zhongguo Fangzhi Daxue Xuebao 21: 17–22
81. Okada A, Ichinose A, Kawanichi S (1996) Polymer 37: 2281–2283
82. Mirzadeh H, Katbab AA, Burford RP (1993) Radiat Phys Chem 42: 53–56
83. Vasilets VN, Kovalchuk AV, Yuranova TI, Ponomarev AN, Talroze RV, Zubarev ER, Plate NA (1996) Polym Adv Technol 7: 173–176
84. Lodge T (1994) Mikrochim Acta 116: 1–31
85. Sheutjens JMHM, Fleer GJ (1980) J Phys Chem 84: 178
86. Zhulina EB, Borisov OV, Pryamitsyn VA, Birshtein TM (1991) Macromolecules 24: 140
87. Zhulina EB, Borisov OV, Pryamitsyn VA (1990) J Colloid Interface Sci 137: 495
88. Shim DFK, Cates ME (1989) J Phys(Paris) 50: 3535
89. Milner ST (1991) Science 251: 905
90. Milner ST, Witten TA, Cates ME (1988) Macromolecules 22: 965
91. Muthukumar M, Ho J (1989) Macromolecules 22: 965
92. Irvine DJ, Mayes AM, Griffith-Cima L (1996) Macromolecules 29: 6037–6043
93. Budkowski A, Klein J, Fetters LJ (1995) Macromolecules 28: 8571–8578
94. Murat M, Grest GS (1989) Macromolecules 22: 4054
95. Grest GS Murat M (1993) Macromolecules 26: 3108–3117
96. Cosgrove T, Heath T, VanLent B, Leemakers F, Scheujens J (1987) Macromolecules 20: 1692
97. Dickman R, Hong DC (1991) J Chem Phys 95: 9288
98. Chakrabarti A, Tiral R (1990) Macromolecules 23: 2016
99. Lai P-Y, Binder K (1991) J Chem Phys 95: 9288
100. Lai P-Y, Binder K (1992) J Chem Phys 97: 586–595
101. Clarke CJ (1996) Polymer 37: 4747–4752
102. Cohen Addad JP, Viallat AM, Pouchelon A (1986) Polymer 27: 843
103. Raudino A, Zuccarello F (1994) J Molecular Structure (Theochem) 314, 125–132
104. Wyart FB, de Gennes PG, Leger L, Marciano Y, Raphaël E (1994) J Phys Chem 98: 9405–9410
105. Aubouy M, Fredrickson GH, Pincus P, Raphaël E (1995) Macromolecules 28: 2979–2981
106. Auvray L, Auroy P, Cruz M (1992) J Phys (Paris) 2: 943
107. Deruelle M, Leger L, Tirrell M (1995) Macromolecules 28: 7419–7428
108. Cosgrove T, Heath TG, Ryan K (1994) Langmuir 10: 3500–3506
109. Janzen J, Song X, Brooks DE (1996) Biophys J 70: 313–320
110. Webber RM, Anderson JL, Jhon MS (1990) Macromolecules 23: 1026–1034
111. Iwata H, Oodate M, Uyama Y, Amemiya H, Ikada Y (1991) J Membrane Sci 55: 119–130
112. Ikeuchi K, Kouchiyama M, Tomita N, Uyama Y, Ikada Y (1996) Wear 199: 197–201
113. Ikada Y, Uyama Y (1993) Lubricating polymer surfaces. Technomic Publ, Lancaster PA
114. de Gennes P-G (1995) C R Acad Sci, Ser 2b 320: 85–89
115. Mukherjee D, Kulkarni A, Gill WN (1994) J Membr Sci 97: 231–249
116. Fytas G, Anastasiadis SJ, Seghrouchini R, Vlassopoulos D, Li J, Factor BJ, Theobald W, Toprakcioglu CT (1996) Science 274: 2041–2044
117. vonGoeler F, Muthukumar M (1995) Macromolecules 28: 6608–6617
118. vonGoeler F, Muthukumar M (1996), J Chem Phys 105: 11335–11346

119. de Gennes PG (1994) C R Acad Sci Paris 318: 165–170
120. Raphaël E, de Gennes PG (1992) J Phys Chem 96: 4002–4007
121. Li Z, Netravali AN (1992) J Appl Polym Sci 44: 333–346
122. Ravichandram V, Obendor SK (1993) In: Mittal KL (ed) Contact angle, wettability, and adhesion. VSP, Utrecht, pp 769–789
123. Penn LS, Jutis B (1989) J Adhesion 30: 67
124. Allmer K, Hult A, Rånby B (1988) J Polym Sci Polym Chem Ed 26: 2099
125. Allmer K, Hult A, Rånby B (1989) J Polym Sci Polym Chem Ed 27: 1641
126. Mori M, Uyama Y, Ikada Y (1994) Polymer 35: 5336–5341
127. Mori M, Uyama Y, Ikada Y (1994) J Polym Sci, Polym Chem 32: 1683–1690
128. Zhang J, Kato K, Uyama Y, Ikada Y (1995) J Polym Sci, Polym Chem 33: 2629–2638
129. Chen K-S, Uyama Y, Ikada Y (1994) Langmuir 10: 1319–1322
130. Kang ET, Neoh KG, Chen W, Tan KL, Liaw DJ, Huang CC (1996) J Adhesion Sci Technol 10: 725–743
131. Wong JY, Kuhl TL, Israelachvili JN, Mulla Nh, Xalipsky S (1997) Science 275: 820–823
132. Zhang J, Uchida E, Suzuki K, Uyama Y, Ikada Y (1996) J Colloid Interface Sci 178: 371–373
133. Zhang J, Uchida E, Uyama Y, Ikada Y (1997) J Colloid Interface Sci 188: 431–438
134. Ikada Y (1994) In: Shalaby SW, Ikada Y, Langer R, Williams J (eds) Interfacial biocompatibility in polymers of biological and biomedical significance. ACS Symposium Series No. 540. American Chemical Society, Washington, D.C., pp 35–48
135. Ikada Y (1994) Biomaterials 15: 725–736
136. Suzuki S, Kishida A, Iwata H, Ikada Y (1986) Macromolecules 19: 1804–1808
137. Uchida E, Uyama Y, Ikada Y (1989) J Polym Sci, Polym Chem 27: 527–537
138. Bergstroem K, Sterberg E, Holmberg K, Hoffman AS, Schuman TP, Kozlowski A, Harris JM (1994) J Biomater Sci Polym Edn 6: 123–132
139. Corretge E, Kishida A, Konishi H, Ikada Y (1988) In: Migliaresi C (ed) Polymers in medicine III. Elsevier Science, Amsterdam, pp.61–72
140. Mok S, Worsfold DJ, Fouda A, Matsuda T (1994) J Appl Polym Sci 51: 193–199
141. Han DK, Jeong SY, Ahn KD, Kim YH, Min BG (1993) J Biomater Sci Polym Edn 4: 579–589
142. Park KD, Lee WK, Yun JY, Han DK, Kim SH, Kim YH, Kim HM, Kim KT (1997) Biomaterials 18:47–51
143. Nakamae K, Miyata T, Matsumoto T (1992) J Membr Sci 69: 121–129
144. Nakamae K, Miyata T, Ootsuki N (1994) Macromol Chem Phys 195: 2663–2575
145. Desai NP, Hubbell JA (1992) Macromolecules 25: 226–232
146. Fujimoto K, Tadokoro H, Ueda Y, Ikada Y (1993) Biomaterials 14: 442–448
147. Fujimoto K, Takebayashi Y, Inoue H, Ikada Y (1993) J Polym Sci Polym Chem 31: 1035–1043
148. Ruckert D, Geuskens G (1995) Eur Polym J 31: 431–435
149. Inoue H, Fujimoto K, Uyama Y, Ikada Y (1997) J Biomed Mater Res 35: 255–264
150. Kishida A, Kato S, Ohmura K, Sugimura K, Akashi M (1996) Biomaterials 17: 1301–1305
151. Uchida E, Uyama Y, Ikada Y (1993) Langmuir 9: 1121–1124
152. Uchida E, Uyama Y, Ikada Y (1994) Langmuir 10: 1193–1198
153. Kulik E, Ikada Y (1996) J Biomed Mater Res 30: 295–304
154. Kato K, Sano S, Ikada Y (1995) Colloids Surf B, Biointerfaces 4: 221–230
155. Fujimoto K, Inoue H, Ikada Y (1993) J Biomed Mater Res 27: 347–355
156. Ivanchenko MI, Kulik EA, Ikada Y (1995) In: Hobett TA, Brash JL (eds) Proteins at interfaces II: fundamentals and applications. ACS Symposium Series No 602, American Chemical Society, Washington DC, pp 463–477
157. Lee JH, Jeong BJ, Lee HB (1997) J Biomed Mater Res 34: 105–114
158. Klee D, Thissen H, Thelen H, Severich B, Hoffmeister K, Von Dahl J, Hanrath P, Höcker H (1996) Proc International Conference on Polymers in Medicine and Surgery, 1–3 July, Glasgow, UK, The Institute of Materials, pp 87–93
159. Ishihara K, Fukumoto K, Aoki J, Nakabayashi N (1992), Biomaterials 13: 229
160. Nakabayashi N, Miyano T, Iwasaki Y, Ishihara K (1996) Proc International Conference on Polymers in Medicine and Surgery, 1–3 July, Glasgow UK. The Institute of Materials, pp 45–51

161. Kang I-K, Kwon OH, Lee YM, Sung YK (1996) Biomaterials 17: 841–847
162. Saito N, Sugawara T, Matsuda T (1996) Macromolecules 29: 313–319
163. Nakamae K, Miyata T, Ootsuki N, Okumura M, Kinomura K (1994) Macromol Chem Phys 195: 1953–1963
164. Kulik EA, Kato K, Ivanchenko MI, Ikada Y (1993) Biomaterials 14: 763–769
165. Wang C-C, Hsiue G-H (1993) J Biomater Sci Polym Edn 4: 357–367
166. Sano S, Kato K, Ikada Y (1993) Biomaterials 14: 817–822
167. Kato K, Ikada Y (1995) Biotechnol Bioeng 47: 557–566
168. Kato K, Ikada Y (1996) Biotechnol Bioeng 51: 581–590
169. Ikada Y, Uyama Y (1993) Lubricating polymer surfaces. Technomic, Lancaster, PA, pp 55–71
170. Uyama Y, Tadokoro H, Ikada Y (1991) Biomaterials 12: 71
171. Inoue H, Uyama Y, Uchida E, Ikada Y (1992) Cell & Mater 2: 21
172. Onishi M, Shimura K (1993) J Photopolymer Sci Technol 6: 421–428
173. Uyama Y, Ikada Y, Tomita N (1993) J Photopolymer Sci Technol 6: 325–330
174. Ikeuchi K, Takii T, Norikane H, Tomita N, Uyama Y, Ikada Y (1993) Wear 161: 179–185
175. Tomita N, Tamai S, Okajima E, Hirao Y, Ikeuchi K, Ikada Y (1994) J Appl Biomater 5: 175–181
176. Okada T, Ikada Y (1995) J Biomater Sci Polym Edn 7: 171–180
177. Okada T, Ikada Y (1991) In: Feng H, Han Y, Huang L (eds) Polymers and biomaterials. Elsevier Science, pp 353–360
178. Nakaoka R, Tsuchiya T, Kato K, Ikada Y, Nakamura A (1997) J Biomed Mater Res 35: 391–397
179. Lee SD, Hsiue GH, Chang PCT, Kao CY (1996) Biomaterials 17: 1599–1608
180. Kato K, Tomita N, Yamada S, Ikada Y (1996) Proc International Conference on Polymers in Medicine and Surgery, 1–3 July, Glasgow UK. The Institute of Materials, pp 255–262
181. Tamada Y, Ikada Y (1994) J Biomed Mater Res 28: 783–789
182. Hsiue GH, Lee SD, Wang CC, Chang PCT (1993) J Biomater Sci Polym Edn 5: 205–220
183. Tretinnikov ON, Kato K, Ikada Y (1994) J Biomed Mater Res 28: 1365–1373
184. Kato K, Eika Y, Ikada Y (1996) J Biomed Mater Res 32: 687–691
185. Yamamoto M, Kato K, Ikada Y (1997) J Biomed Mater Res 37: 29–36
186. Kamei S, Tomita N, Tamai S, Kato K, Ikada Y (1997) J Biomed Mater Res 37: 384–393

Editor: Prof. T. Saegusa
Received: June 1997

Phase Structure of Polyethylene and Other Crystalline Polymers by Solid-State ^{13}C NMR

Ryozo Kitamaru

11 Hanazono Enjyoji, U Kyoku, Kyoto 6168027, Japan. E-mail: rk11@mba.infosphere.or.jp

1	Introduction	42
2	Solid-State ^{13}C NMR	43
2.1	High-Resolution ^{13}C NMR Spectrum	43
2.2	Spin Relaxation	46
3	Linear Polyethylene Crystallized from the Melt	48
3.1	Introduction and Approach by Broad-Line ^{1}H NMR	48
3.2	Approach by High-Resolution ^{13}C NMR	51
3.3	Discussion	59
4	Linear Polyethylene Crystallized from Dilute Solution	61
4.1	Introduction and Approach by Broad-Line ^{1}H NMR	61
4.2	Approach by High-Resolution ^{13}C NMR	62
4.3	Discussion	63
5	High-Pressure Crystallized Polyethylene	64
5.1	Introduction	64
5.2	Experimental	65
5.3	Discussion	69
6	Ultra High Modulus Polyethylene	70
6.1	Introduction	70
6.2	Experimental	70
6.3	Phase Structure and Discussion	72
7	Linear Polyethylene with Randomly Distributed Ethyl Branches (Hydrogenated Polybutadiene)	74
7.1	Introduction	74
7.2	Phase Structure as Revealed from the Analysis of Resonance Lines of Main Chain Methylene Carbons	75
7.3	Partitioning of Methyl Carbons Between Different Phases	77

8	Poly(tetramethylene oxide)	79
8.1	Introduction	79
8.2	Experimental	80
8.3	Phase Structure and Discussion	81
9	**Isotactic Polypropylene Crystallized from the Melt**	84
9.1	Introduction	84
9.2	Experimental	85
9.3	Phase Structure and Discussion	87
10	**Syndiotactic Polypropylene Gel**	89
10.1	Introduction	89
10.2	Experimental and Discussion	90
11	**Concluding Remarks**	98
	References	99

1
Introduction

Crystalline polymers generally comprise a variety of phase structures, including the crystalline and noncrystalline components. The noncrystalline component is thought to be in a supercooled state of the molten state of polymers, either in the rubbery or glassy state. The question is whether a somewhat ordered noncrystalline component exists or not, due to the coexistence with the crystalline component. Flory reported in 1949 that the boundary between the crystalline and amorphous regions of most long-chain molecules will not be well-defined as is typical of monomeric systems [1]. Subsequent theoretical analyses, involving several methods, have quantitatively established the existence of an interfacial region that comprises a transition phase from the crystalline to the amorphous phase [2–11]. Over the past few years, a variety of experimental methods have confirmed the expectation of such an interphase [12]. These methods involve broad-line ^{1}H NMR [13–16] high-resolution solid-state ^{13}C NMR [17–31] Raman spectroscopy [32–35], and small-angle X-ray and neutron scattering, among others [36–42].

To elucidate the phase structure in detail it is necessary to characterize the molecular chain conformation and dynamics in each phase. However, it is rather difficult to obtain such molecular information, particularly of the noncrystalline component, because it is substantially amorphous. In early research in this field, broad-line ^{1}H NMR analysis showed that linear polyethylene crystallized from the melt comprises three components with different molecular mobilities; solid, liquid-like and intermediate molecular mobility [13–16]. The solid component was attributed to molecules in the crystalline region, the liquid component to that in the noncrystalline amorphous and the intermediate component to that in an intermediate region between the crystalline and amorphous regions. A similar approach was also used on samples crystallized from dilute solution and some characteristics of the phase structure of the samples was elucidated. However, since broad-line ^{1}H NMR can characterize substances only by overall molecular mobility, further detailed information with regard to molecular chain conformation and alignment in each component could not be obtained. Hence, some objection to the broad-line ^{1}H NMR analysis could remain [49].

Rather recently, we have studied the solid-state structure of various polymers, such as polyethylene crystallized under different conditions [17–21], poly (tetramethylene oxide) [22], polyvinyl alcohol [23], isotactic and syndiotactic polypropylene [24, 25], cellulose [26–30], and amylose [31] with solid-state high-resolution ^{13}C NMR with supplementary use of other methods, such as X-ray diffraction and IR spectroscopy. Through these studies, the high resolution solid-state ^{13}C NMR has proved very powerful for elucidating the solid-state structure of polymers in order of molecules, that is, in terms of molecular chain conformation and dynamics, not only on the crystalline component but also on the noncrystalline components via the chemical shift and magnetic relaxation. In this chapter we will review briefly these studies, focusing particular attention on the molecular chain conformation and dynamics in the crystalline-amorphous interfacial region.

2
Solid-State ^{13}C NMR [43]

2.1
High-Resolution ^{13}C NMR Spectrum

In this chapter we review the study of the solid structure of many crystalline polymers mainly with high-resolution solid-state ^{13}C NMR, so we will briefly summarize here its principles for the convenience of the reader. For this purpose we first consider the Hamiltonian of an ensemble of nuclei possessing spin in a static field $\mathbf{B}_0$. The Hamiltonian to be considered of this spin system can be written as:

$$H = H_Z + H_S + H_D + H_J$$

Here H_Z represents the direct interaction energy of spins with the external magnetic field $\mathbf{B}_0$. This term, the so-called Zeeman energy, gives the main resonance line as a δ-function for each nucleus with spin. H_S, chemical shift Hamiltonian, describes the indirect interaction of spins with $\mathbf{B}_0$ via electrons, and gives different shifts to the main resonance peak that can distinguish each constituent nucleus in a molecule by difference of the electromagnetic environment. H_D and H_J describe the direct and indirect (via electrons) dipolar interactions between spins, respectively. The former gives the line width to the main resonance line and the latter, the so-called J-coupling or spin-coupling, brings splitting or satellites. In addition to these terms there are some other terms, e.g. quadrupole Hamiltonian. In this chapter we only consider ^{13}C NMR for spin systems whose nuclei have spins of either zero or 1/2 and, therefore, such further terms are omitted.

In ^{1}H or ^{13}C NMR measurements in solution, the direct dipolar interaction H_D actually disappears because, due to rapid molecular motion, the interspin (internuclear) vectors are rapidly space-averaged within the time scale of the measurements. Hence, H_S and H_J are detectable as sharp lines or splittings in a high-resolution ^{1}H or ^{13}C spectrum and they can be related to the detailed molecular structure or conformation of the substance being investigated. In a solid, however, the directions of the internuclear vectors are stationarily fixed in time even if they are distributed randomly in space. Then, H_D gives a very wide line width to the main resonance line and completely masks all lines due to H_S and H_J. Therefore, in usual NMR measurements, one observes only a very broad resonance spectrum which is determined by H_Z and H_D. Some information is, of course, obtainable, even if the spectrum is very broad, because the line shape that is determined by H_D reflects the geometrical alignment of nuclei and the line width reflects the degree of molecular motion. However, a spectrum in which the terms due to H_S are detectable cannot be obtained and, therefore, is termed the broad-line NMR spectrum.

To obtain high-resolution spectrum for solid matter, we have to eliminate the effect of H_D and furthermore average that of H_S, because H_S is a tensor of rank 2. In the solid-state ^{13}C NMR, the high-resolution spectrum is obtained by combining three techniques, ^{1}H dipolar decoupling (DD), magic angle spinning (MAS), and cross-polarization (CP). DD is a technique to eliminate the effect of H_D, which is achieved by applying a resonant oscillating field $\mathbf{B}_1$ of sufficient amplitude on ^{1}H spins. If the resonant oscillating field is applied, the dipolar field between ^{13}C and ^{1}H and between ^{1}H spins themselves now changes the direction rapidly, so that the dipolar interactions H_D between ^{13}C and ^{1}H, and between ^{1}H spins themselves, are removed. Furthermore, the interaction between ^{13}C nuclei is negligible because of their low natural abundance (ca. 1.1%). Hence the ^{13}C magnetization, without any contribution from H_D, can be detected. With use of ^{1}H dipolar decoupling (DD), the spectrum which contains only the contribution from H_Z and H_S can be obtained. The spectrum thus obtained can be called the H_S spectrum, i.e. chemical shift spectrum. However, high-resolution spectrum still cannot be obtained in general, because the chemical shift Hamiltonian H_S has anisotropy as a tensor quantity of rank 2 as mentioned above. The elimination of this anisotropy is usually achieved by rotating samples rapidly around an axis inclined at an angle of 54° 44' to the static field $\mathbf{B}_0$ during measurement. Here 54° 44' is the so-called magic angle that fulfills the relationship:

$$1 - 3\cos^2\theta = 0.$$

This process is called magic angle spinning (MAS). If the rate of spinning is larger than the line width caused by the anisotropy of H_S in the frequency scale, the anisotropy is removed and a sharp resonance line appears at the average of the principal values of H_S as a tensor quantity. Thus, a high-resolution ^{13}C NMR spectrum can be obtained by using two procedures, DD and MAS.

In addition to MAS and DD explained above, cross polarization (CP) is usually used in present solid-state high-resolution ^{13}C NMR measurements. This is a process to enhance ^{13}C magnetization by cross polarization (CP) between ^{13}C and ^{1}H spins. The ^{13}C magnetization is enhanced by the equilibrium magnetization of abundant ^{1}H spins. This procedure is achieved by contacting the ^{13}C spin system to the ^{1}H system while Zeeman energies of both ^{13}C and ^{1}H spins against the respective sub-field B_1 are equalized. A ^{13}C magnetization as large as 4 times its equilibrium magnetization is obtained by this process. Furthermore, in this case, the time demanded for one measuring pulse sequence is determined by the spin-lattice relaxation time T_{1H} of ^{1}H that is usually much shorter than the T_{1C} of ^{13}C. Hence, the time required for obtaining a spectrum with a sufficient signal-noise ratio can be largely shortened by a factor of 1/100 or sometimes 1/1000 by using a CP in comparison to the measurements without CP. However, it should be noted that the CP efficiency may differ according to the molecular mobility of samples. Sometimes, we could not use CP for the purpose of characterizing different phases with different molecular mobilities in samples.

In Fig. 1, the pulse sequences frequently used in this work are schematically shown. The pulse sequence I is used for obtaining DD/MAS ^{13}C NMR spectrum without the CP procedure. The ^{13}C magnetization which appears in the B_0 direction for a time of τ_ℓ is turned through 90° by a resonant field and the free induc-

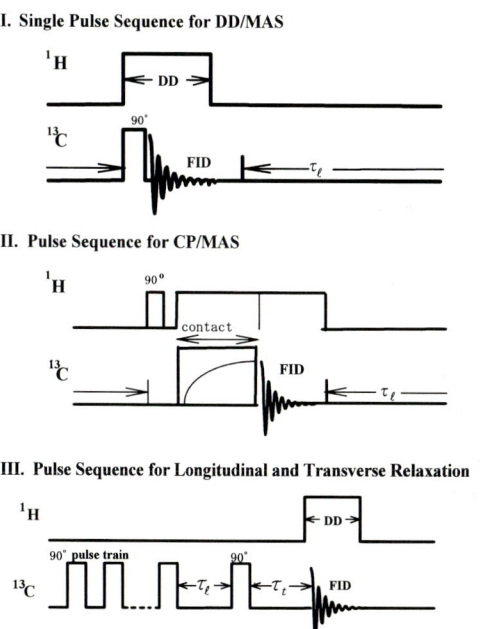

Fig. 1. Pulse sequences for high-resolution ^{13}C NMR. **I.** Single pulse sequence for DD/MAS without CP. **II.** Pulse sequence for CP/MAS. **III.** Pulse sequence for longitudinal and transverse relaxation. See the text for detail

tion decay (FID) is observed in the xy plane (perpendicular to B_0) under ^{1}H decoupling (DD) with MAS throughout the measurement. Here, if the waiting time τ_ℓ, that is actually the repetition time of the pulse sequence, is set to be as long as 5 times the T_{1C}, thermal equilibrium spectrum can be obtained [44]. Here we sometimes express this pulse sequence as (90°–FID$_{DD}$– τ_ℓ)$_n$, where n is the repetition number of the measurement. Sequence II is that required to obtain the CP/MAS ^{13}C NMR spectrum. The ^{1}H magnetization which appears in the B_0 direction for a time of τ_ℓ is turned through 90° and locked in the ^{1}H resonant field and the ^{13}C magnetization obtained by the cross polarization with the ^{1}H system is observed under DD. The cross polarization is performed by contacting both spin systems by applying the ^{13}C resonant field while Zeeman energies of both ^{1}H and ^{13}C spin systems in relation to respective resonant fields are equalized.

In addition to pulse sequences I and II, various pulse sequences were used in this work. Figure 1, shows a pulse sequence for obtaining the partially relaxed spectrum in longitudinal and transverse directions (Pulse sequence III). After saturating by a 90° pulse train, the ^{13}C magnetization appearing longitudinally is turned 90° and allowed transverse relaxation for τ_t and the FID is observed under DD. Using this pulse sequence, the longitudinal and transverse relaxations of the different components, particularly shorter T_{1C} components in polymers, can be examined.

2.2
Spin Relaxation

High-resolution ^{13}C NMR spectra give detailed information of molecular alignment of polymers via chemical shift that distinguishes each carbon in the molecule. Furthermore, studies of the ^{13}C magnetic relaxation provide information about molecular motion with relation to each carbon. In the case of organic substances and polymers that contain only ^{1}H and ^{13}C as spin having nuclei, the relaxation is conducted by the time fluctuation of the magnetic dipolar interactions between ^{13}C and ^{1}H spins. Hence, it provides information about molecular motion via the time fluctuation of interspin vector (i.e. internuclear vector), connecting ^{13}C and ^{1}H nuclei.

The relaxations in the directions parallel and perpendicular to B_0, i.e. the longitudinal and transverse relaxations, are often measured in present ^{13}C NMR. The relaxation parallel to B_0 is called the longitudinal relaxation or spin-lattice relaxation, or simply T_1 relaxation, since it is carried out by energy exchange between the spin system and the lattice. On the other hand, the relaxation perpendicular to B_0 is called transverse relaxation or spin-spin relaxation, or simply T_2 relaxation, since it is conducted by the spin exchange inside the spin system. Hereafter, we designate the spin-lattice and spin-spin relaxations of the ^{13}C magnetization as T_{1C} and T_{2C}, and those of ^{1}H as T_{1H} and T_{2H}, respectively. These are given as functions of the spectral density which is defined as the Fourier transform of the time-dependent part of the dipolar interaction. Since this time-dependent part is a function of the orientation of the internuclear vector, it is generally called the orientation function. That is T_{1C} and T_{2C} are given by the spectral densities, J $(\omega)_m$, that are defined as Fourier transforms of the correlation functions of the different terms of the orientation function, and can be defined as in Eqs. (1) and (2), respectively.

$$1/T_{1C} = (1/20)K\,[J_0(\omega_H-\omega_C) + 3J_1(\omega_C) + 6J_2(\omega_H+\omega_C)] \tag{1}$$

$$1/T_{2C} = (1/40)K\,[4J_0(0) + J_0(\omega_H-\omega_C) + 6J_1(\omega_H) + 3J_1(\omega_C) + 6J_2(\omega_H+\omega_C)] \tag{2}$$

Here the spectral densities are defined as:

$$J(\omega)_m = \int_{-\infty}^{\infty} \overline{F_m^*(\tau)\,F_m(0)}\,\exp(-i\omega\tau)d\tau / |\overline{F_m(0)^2}|$$

$F_m(t)$ are different terms, designated by suffix m, of the orientation function in the dipolar interaction and $\overline{F_m^*(\tau)\,F_m(0)}$ are their correlation functions. ω_H and ω_C are the resonant frequencies (Larmor frequencies) of ^1H and ^{13}C. They are defined by magnetogyric ratios, γ_H and γ_C, of ^1H and ^{13}C and the field intensity B_0 as $\omega_H = -\gamma_H B_0$, $\omega_C = -\gamma_C B_0$.

K is a constant defined as

$$K = (\mu_0/4\pi)^2\,\gamma_H^2\gamma_C^2\hbar r^{-6}$$

where μ_0 is the magnetic transmittance of vacuum and $\hbar$ Planck's constant/2π. r is the internuclear distance between ^{13}C and ^1H. If the correlation functions evolve exponentially with one time constant as τ_c as

$$\overline{F_m^*(\tau)\,F_m(0)} = |\overline{F(0)^2}|\,\exp(-|\tau|/\tau_C) \tag{3}$$

the longitudinal and transverse relaxation times are given as

$$1/T_{1C} = (1/20)K\left[\frac{2\tau_C}{1+(\omega_H-\omega_C)^2\tau_C^2} + \frac{6\tau_C}{1+\omega_C^2\tau_C^2} + \frac{12\tau_C}{1+(\omega_H+\omega_C)^2\tau_C^2}\right] \tag{4}$$

$$1/T_{2C} = (1/40)K\left[8\tau_C + \frac{2\tau_C}{1+(\omega_H-\omega_C)^2\tau_C^2} + \frac{12\tau_C}{1+\omega_H^2\tau_C^2}\right.$$

$$\left. + \frac{6\tau_C}{1+\omega_C^2\tau_C^2} + \frac{12\tau_C}{1+(\omega_H+\omega_C)^2\tau_C^2}\right] \tag{5}$$

since the Fourier transform of the correlation functions is given as

$$\int_{-\infty}^{\infty} \exp(-|\tau|/\tau_C)\exp(-i\omega\tau)d\tau = \frac{2\tau_C}{1+\omega^2\tau_C^2}$$

where τ_c is the relaxation time of the motion. The theory with relation to the motion dictated by only one relaxation time (correlation time) is called single correlation-time theory.

As revealed from Eqs. (1) and (2), or their candid forms (4) and (5), the longitudinal relaxation is determined by the spectral densities in the order of ω_H ω_C, whereas the transverse relaxation involves the contribution from the zero frequency component $J_0(0)$. In the case of solid matter, τ_C is generally very long. Hence, the transverse relaxation is predominantly determined by the zero frequency component $J_0(0)$. In Eq. (5), for example, the zero frequency term (the first term) dominates the other terms that are reciprocally proportional to τ_C for $\omega^2\tau^2 \gg 1$. T_{1C} increases as τ_C increases (i.e. as the material under consideration becomes solider), whereas T_{2C} decreases infinitely as τ_C increases. For example, T_{1C} is generally in an order of several tens ~ several hundreds of seconds for the crystalline component and in an order of a few tenths of a second for rubbery components of polymers. On the other hand, T_{2C} is of an order of a few tens of microseconds for the crystalline or glassy component and a few milliseconds for the rubbery component of polymers. In this work, T_{1C} and T_{2C} are used for characterizing different components in crystalline polymers.

3
Linear Polyethylene Crystallized from the Melt

3.1
Introduction and Approach by Broad-Line ^{1}H NMR

Linear polyethylene in the solid-state has been most extensively studied as a typical representative of a crystalline polymer by various methods because of the simplicity of its molecular and crystal structure. Figure 2-A shows a broad-line ^{1}H NMR spectrum in the first derivative form of linear polyethylene at 20 °C. The sample is a molecular weight fraction with a viscosity-averaged molecular weight of 90,000, which was crystallized isothermally at 130 °C from the melt. As pointed out in the last section, one can only see in this broad-line ^{1}H NMR spectrum the contribution of the dipolar interaction H_D between the ^{1}H spins themselves, since all contributions from H_S or H_J that provide detailed information about the molecular alignment in the sample are masked by H_D. However, the broad line shape narrows and approaches a Lorentzian function as the molecular mobility increases. Hence, one can characterize the structure of the sample by examining the line shape and line width in terms of molecular mobility. At first glance the sample may comprise two different components with different line widths, one being a rather broad Gaussian and the other a narrow Lorentzian function, that may correspond to the crystalline and amorphous phases, respectively. However, if this kind of spectrum is analyzed into two components, assuming a two-phase structure of crystalline and amorphous phases as shown in Fig. 2-B, (either straight line decomposition [45] or symmetrical decomposition [46, 47] the broad component exceeds the crystalline fraction determined by X-ray diffraction analysis or density measurements.

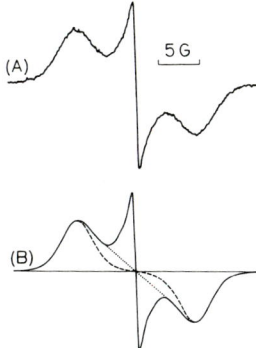

Fig. 2. a 60 MHz broad-line ¹H NMR spectrum of bulk polyethylene at 20 °C. **b** Two-component analyses, *dotted line:* straight line decomposition [45] *dashed line:* symmetrical decomposition [46, 47]

This suggests that the broad component includes some contribution from a rigid noncrystalline component.

Bergmann et al. broke down such spectra for polyethylene into three components, i.e. broad, intermediate, and narrow components [13, 14]. The methylene groups of the polymer were divided into three classes: rigid, hindered-rotational, and micro-Brownian mobile methylene groups, and these were considered to contribute to the broad, medium, and narrow components, respectively. The mass fraction of the broad component thus estimated was in good accord with the crystalline fraction obtained from X-ray diffraction analysis and density measurements. The hindered-rotational and micro-Brownian mobile methylene groups were assigned to two noncrystalline phases, the former with less mobility and the latter with a liquid-like mobility. The relaxation processes, called α, β and γ, observed in the dielectric and mechanical measurements were well explained as corresponding to the processes associated with a local molecular movement in the crystalline region, and micro-Brownian and limited molecular motions in the noncrystalline region, respectively.

We have examined linear polyethylene samples, covering a very wide range of molecular weight, that were crystallized from the melt by a similar three-component analysis of the broad-line ¹H NMR spectra and obtained very important information about the phase structure of the samples [15, 16]. Figure 3 shows the molecular-weight dependence of the mass fractions of the three components at room temperature. As can be seen, although the narrow component is negligible for samples of molecular weights smaller than 31,800 it appears clearly at a molecular weight of 44,900 and increases as the molecular weight increases. The mass fraction of the medium component is as small as 0.05–0.08 for samples whose narrow component is actually devoid but it gradually increases with the appearance of the narrow component and reaches a rather high level for samples with higher molecular weights. In other words, the samples with very low molecular weight are only made up of crystalline lamellae and noncrystalline interlamellar material without a liquid-like amorphous component. The amorphous

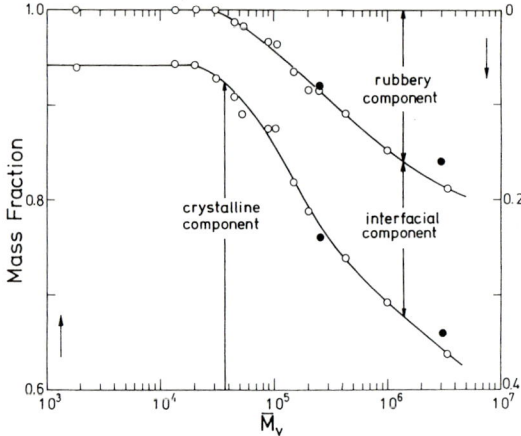

Fig. 3. Mass fractions of crystalline, rubbery, and crystalline-amorphous interfacial phases of bulk polyethylene as a function of molecular weight $\overline{M}_v$. ○: data by broad-line ^{1}H NMR analysis. ●: data by high-resolution ^{13}C NMR analysis

component appears after the mass fraction of the interlamellar material reaches a chosen level.

These studies provide important information about the interlamellar material of crystalline polyethylene as a function of molecular weight. The samples with molecular weight less than about 30,000 are composed of the crystalline and interfacial regions without an interzonal region with liquid-like character. Within this molecular weight range, the average ratio of the lamellar thickness to the extended molecular chain length, ζ/x, changes from unity to about 1/4 [48]. Hence, a molecular chain will participate in a crystal lamella almost always only once, or at least less than three times, so that molecular chain folding will be very scarce in forming the lamellar structure. The noncrystalline molecular chains that are excluded from the crystalline region will be somewhat severely restricted by the crystalline molecular chains in conformation and mobility. The non-crystalline component with random conformation with liquid-like mobility will not appear.

As the molecular weight increases from 45,000 to 100,000, the ratio ζ/x varies from 1/4 to 1/7. The number of molecular chains that penetrate several crystal lamellae, or participate repeatedly in a crystal lamella by molecular chain folding, will gradually increase. The amorphous region, in which the molecular conformations are distributed over all allowable conformation and rapidly transformed each other, can appear in this molecular weight range. The transition region from the crystalline to this amorphous region (the crystalline-amorphous interfacial region) will comprise molecular chains whose conformation and motion are severely restricted, including folded molecular chains, entanglement and molecular end groups. However, since broad-line ^{1}H NMR can characterize the phase structure only in terms of the molecular mobility through the line shape and line width or the second moment of the spectrum, it is difficult to obtain further detailed information about molecular alignment or conformation and hence some objection to our analysis could remain [49]. However, we were able to obtain more detailed information about the phase structure, not only for these samples but also for many crystalline polymers, by use of solid-state high-resolution ^{13}C NMR [17–31]. We will discuss the above-cited results obtained by ^{13}C NMR later.

3.2
Approach by High-Resolution ^{13}C NMR

High-Resolution ^{13}C NMR Equilibrium Spectrum. Figure 4 shows the DD/MAS ^{13}C NMR spectrum at room temperature for a linear polyethylene of molecular weight of 3.0×10^6 that was crystallized isothermally at 130 °C from the melt. This spectrum was obtained at a field strength of 4.7 T (Larmor frequency of ^{13}C, $\omega_c = 50$ MHz) by a single pulse sequence ($\pi/2$ - FID$_{DD}$ - 1700s)$_n$ that is schematically shown in Fig. 1–I. In this pulse sequence, since τ_ℓ was set at 1700 s longer than 5 times the longest T_{1C} of this sample, the spectrum expresses the thermal equilibrium state. In the spectrum two distinct resonance lines can be recognized at 33 and 31 ppm relative to the chemical shift of tetramethylsilane (TMS); designated hereafter as resonance lines I and II, respectively. Since the 33 ppm value corresponds to the average of the principal values of the chemical shift tensor for *trans-trans* methylene sequences of polyethylene or n-paraffins in orthorhombic crystalline form [50–52] and the 31 ppm value is close to that observed for polyethylene in solution, we simply assign these lines to the orthorhombic crystalline and the noncrystalline component, respectively. This spectrum expresses a thermal equilibrium state as mentioned above. Hence, it reflects strictly the contribution from all components in the sample. However, since simple examination of this spectrum does not provide further information, we have examined the relaxation phenomenon of each line.

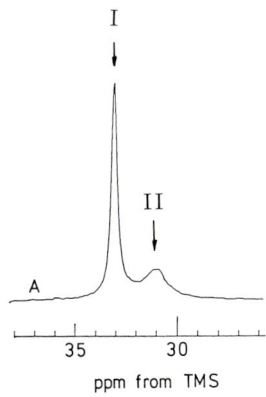

Fig. 4. 50 MHz DD/MAS ^{13}C NMR spectrum of bulk polyethylene with a viscosity-average molecular weight of 3.0×10^6 at room temperature. The spectrum was obtained by pulse sequence I with a repetition time, $\tau_\ell = 17,000$ s. The chemicalshift is based on that of tetramethylsilane (TMS)

Spin-Lattice Relaxation. In order to determine whether each resonance line comprises a single component, we first measured the spin-lattice relaxation time T_{1C} by the pulse sequence developed by Torchia [53] or by the standard saturation-recovery pulse sequence. The T_{1C} values thus obtained were 2560, 263 and 1.7 s for resonance line I and 0.37 s for line II. As reported by several investigators, the line at 33 ppm is associated with three different T_{1C} values [17, 54, 55]. This means that this line is contributed to by three components with different molecular mobilities. However, since each component was represented by a single Lorentzian line shape at 33 ppm, they are all assignable to methylene groups in the orthorhombic crystalline form or in the *trans-trans* conformation. The component with a T_{1C} of 1.7 s can be assigned to methylene groups with a somewhat pronounced molecular motion in the vicinity of the crystalline stem end. Two longer T_{1C}'s, 2560 and 263 s, can be assigned to thicker and thinner lamellar crystalline components. On the other hand, the noncrystalline component which appears at 31 ppm is associated with a single T_{1C} of 0.37 s. This simply implies that the noncrystalline component comprises a single phase in as much as judged only by T_{1C}.

In order to determine the content of this noncrystalline line further, we examined in more detail the behavior of the spin-lattice relaxation. Figure 5 shows the partially relaxed spectra in the course of the inversion recovery pulse sequence $(180°-\tau-90°-FID_{DD}-10s)_{120}$ with varying τ values. The magnetization that was recovered for 10 s in the z direction was turned to negative z direction by 180° pulse and the magnetization recovered in z direction for varying τ was measured in the xy plane under ^1H DD. The spectra at different steps of the longitudinal relaxation were obtained by Fourier transform and are shown in Fig. 5. In these spectra the contribution from the crystalline components with T_{1C}'s of 2,560 and 263 s are eliminated due to the lack of time for recovery at each pulse sequence. Therefore, we observed preferentially the relaxation process of the noncrys-

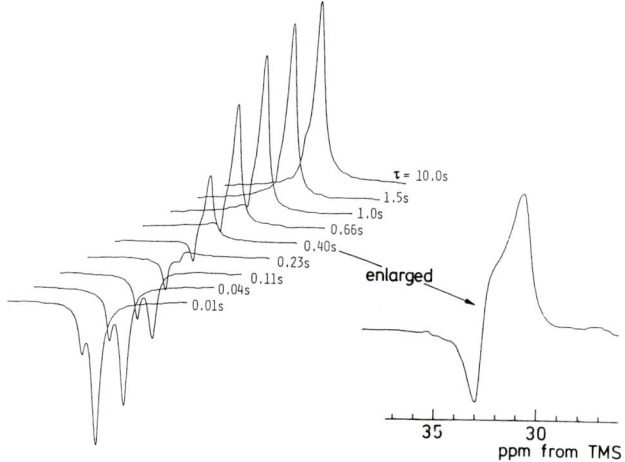

Fig. 5. Partially relaxed spectra of the bulk polyethylene with $\overline{M}_v$ of 3.0×10^6, taken by inversion recovery pulse sequence $(180°-\tau-90°-10s)_{120}$ with different relaxation times τ's

talline component at 31 ppm. See the spectrum at τ = 0.4 s, an enlargement of which is shown at the right Fig. 5. The line at 33 ppm due to the crystalline component with T_{1C} of 1.7 s is still in the negative z direction, while the line at 31 ppm due to the noncrystalline component has already appeared in the positive z direction. It is to be noted here that a downfield shoulder is visible in the line at 31 ppm. This suggests the presence of another line due to the noncrystalline component with a different chemical shift. Although the noncrystalline component comprises a single phase judged only by T_{1C}, two noncrystalline components with the same T_{1C} and with somewhat different chemical shifts must be assumed. Since the two noncrystalline phases assumed here are associated with the same T_{1C} value, they should have the same molecular mobility in the T_{1C} relaxation time frame but not necessarily the same in the other relaxation time frames because of different molecular chain conformation (different chemical shift).

Spin-Spin Relaxation. In order to determine the content of these two noncrystalline components, we next examined the transverse relaxation by a pulse sequence shown schematically in Fig. 1–III. The partially recovered magnetization in the z direction for a $τ_ℓ$ of 3.5 s was followed by transverse relaxation for varying time $τ_t$, and the FID was observed under ^{1}H DD. The spectra at different steps of transverse relaxation were thus obtained, and the result is shown in Fig. 6. The spectrum at $τ_t$ = 0.5 μs (a dead time in this pulse sequence) is assumed

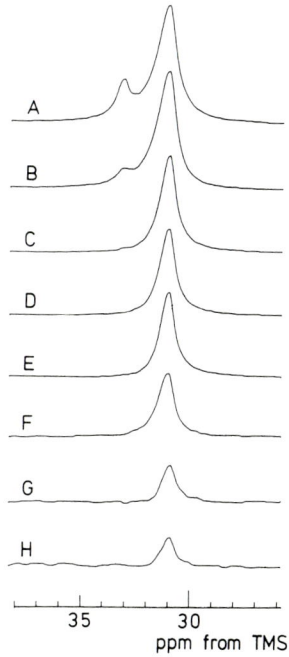

Fig. 6. Partially relaxed spectra of the bulk polyethylene with $\overline{M}_v$ of 3×10^6, taken by pulse sequence II with $τ_ℓ$ = 3.5 s. A, B, C, D, E, F, G, and H show the transversally relaxed spectra for $τ_t$ = 0.5, 20, 60, 100, 140, 500, 2000, and 4000 μs, respectively

to contain the contributions from all noncrystalline components and a part of the crystalline component with $T_{1C} = 0.7$ s. As is seen, with increasing τ_t the contributions from the crystalline and noncrystalline components with shorter transverse relaxation times quickly disappear, leaving only the contribution from the noncrystalline component with a longer transverse relaxation time. Thus, at $\tau_t = 100$ μs only a single peak remains at 31.0 ppm. With further increasing τ_t, this single line gradually decreases in intensity but the line shape remains essentially unchanged.

The peak height of the noncrystalline resonance line at 31 ppm during the transverse relaxation is plotted against the relaxation time τ_t in Fig. 7. It can be seen that the overall decay curve (the bottom line in the figure) can be clearly resolved into two parts, a rapid decay within 50 μs and a subsequent slow decay. The initial slope of the slow decay yielded $T_{2C} = 2.4$ ms, and the T_{2C} of the rapid decay was estimated to be 44 μs by the usual decay analysis as shown at the top of the figure. This indicates that the two noncrystalline components are associated with T_{2C}'s of 44 μs and 2.4 ms. Together with the longitudinal relaxation phenomena shown in Fig. 5, it can be concluded that the 31 ppm resonance comprises two lines, a downfield line with T_{2C} of 44 μs and an upfield line with T_{2C} of 2.4 ms, both associated with the same T_{1C} of 0.37 s. The latter line can be assigned to the amorphous component in the rubbery state, since a T_{2C} of 2.4 ms is typical of rubbery polymers. On the other hand, the downfield line can be assigned to a noncrystalline component whose molecular motion is severely restricted, although its motion in the T_{1C} time frame is the same as the rubbery amorphous component.

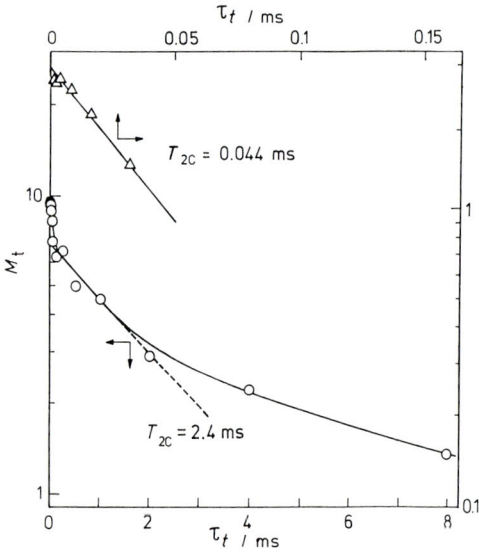

Fig. 7. Transverse relaxation of 31 ppm resonance line: the peak height at 31 ppm in Fig. 6 is plotted against transverse relaxation time τ_1

Line Shape Analysis of Thermal Equilibrium Spectrum. It has been shown that the crystalline region comprises three components with different T_{1C} values and the noncrystalline region consists of two regions with different molecular mobilities, i.e. rubbery and rigid noncrystalline components. Based on this fact we analyzed the thermal equilibrium spectrum shown in Fig. 4. For this purpose, we first determined the elementary line shape of each component. Firstly it was found that the elementary line shape of the crystalline component can be taken as a Lorentzian function at 33 ppm. The elementary line shapes of the two noncrystalline components at ca. 31 ppm were obtained by analyzing the actual spectra during the transverse relaxation shown in Fig. 6; the process is shown in Fig. 8. Here A is a longitudinally relaxed spectrum for τ_ℓ of 3.5 s (spectrum A at $\tau_t = 0.5$ µs in Fig. 6), which comprises the contribution of a crystalline component with T_{1C} of 1.7 s and full contribution from the two noncrystalline components. The elementary line shape of the rubbery amorphous component was obtained from the spectra after the contributions from the crystalline component and the rigid noncrystalline component with T_{2C} of 44 µs had disappeared. As mentioned above, the partially relaxed spectra at $\tau_t > 100$ µs have an essentially unchanged line shape at 31 ppm, typical of rubbery polymers. Hence, the partially relaxed spectrum at $\tau_t = 140$ µs was taken as the elementary line shape of the amorphous noncrystalline component (B in Fig. 8). Spectrum C is a difference spectrum that was obtained by subtracting B from A. It comprises the contribution from the crystalline component with T_{1C} of 1.7 s and the rigid noncrystalline component. Since spectrum C can be broken down well into two Lorentzian line shapes at 33.0 and 31.3 ppm, the latter line shape at 31.3 ppm was taken as the elementary line shape of the rigid noncrystalline components. Thus all elementary line shapes of the crystalline and two noncrystalline components were determined, and were found to be well approximated by Lorentzian line shapes with different line widths.

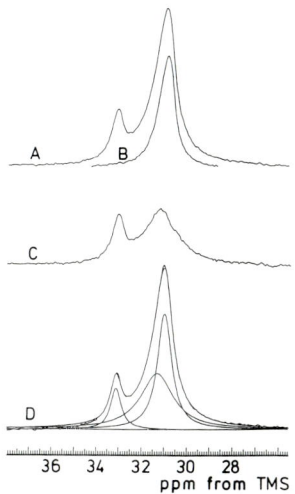

Fig. 8. Line shape analysis of the noncrystalline resonance of the bulk polyethylene with $\overline{M}_v$ of 3×10^6. A and B correspond to A and E in Fig. 6, respectively, C is the difference spectrum (A–B), and D shows the line shape analysis. In D, the composite curve of the component lines is mostly superimposed on the experimental spectrum

On the basis of the elementary line shapes thus obtained, spectrum A has been analyzed in terms of three Lorentzian lines by least-squares fitting; the line width and the peak height of each component have been varied while keeping the peak positions of 33.0, 31.3, and 31.0 ppm. As shown in Fig. 8D, the composite curve of the three Lorentzian functions agrees well with the observed spectrum A. Based on this analysis, the total thermal equilibrium spectrum has been analyzed into three Lorentzian functions centered at 33.0, 31.3, and 31.0 ppm with use of the elementary line shapes thus obtained. Although the crystalline line at 33.0 ppm comprises three components with different T_{1C}'s, the line is assumed to be a single Lorentzian. The result is shown in Fig. 9, where the integrated intensity fractions of the respective components are also shown. As can be seen, the composite curve indicated by a dotted line reproduces well the observed spectrum. The integrated intensity fraction of the crystalline line is in good accord with the degree of crystallinity estimated by the broad-line ^{1}H NMR analysis; the latter also coincides with the value obtained from density measurements. Therefore, the result obtained here confirms the conclusion previously obtained by broad-line ^{1}H NMR analysis, in terms of chemical shift and relaxation phenomena, that the linear polyethylene consists of a crystalline component and two noncrystalline components with different molecular mobilities.

The half-width of the crystalline component line was estimated to be 18 Hz. This value reflects the very stable orthorhombic crystalline phase of this sample. The component line shape centered at 31.0 ppm represents the contribution from the amorphous phase in which the molecular conformations are changed rapidly over all permitted conformations. The relatively narrow line width estimated as 38 Hz is caused by the rapid molecular motion. The line centered at 31.3 ppm represents the noncrystalline phase in which the local molecular motion can occur in the same manner as in the amorphous phase (in T_{1C} time frame), but a long-range molecular motion accompanying a conformational change related to a 10–20 methylene sequence is severely restricted. The wide line width as 85 Hz

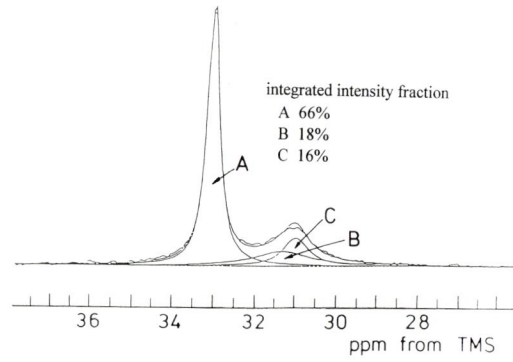

Fig. 9. Line shape analysis of the equilibrium ^{13}C NMR spectrum of bulk polyethylene shown in Fig. 4. A: crystalline component centered at 33.0 ppm, B: crystalline-amorphous interfacial component at 31.3 ppm, C: amorphous component at 31.0 ppm. The composite curve of the component lines is mostly superimposed on the experimental spectrum

indicates that the molecular conformation is mostly fixed over all permitted conformations. Considering the wide symmetric lineshape and the chemical shift, it can be concluded that this phase comprises the transition region from the crystalline to the amorphous phase.

In addition to T_{1C} and T_{2C}, the proton spin-lattice relaxation time, T_{1H}, was also measured by observing the ^{13}C magnetization under ^1H DD that was obtained by a crosspolarization from ^1H magnetization that was recovered in an inversion recovery process. The T_{1H} provides not only the proper conditions for the CP/MAS measurements of ^{13}C NMR but also the information about the location of different phases via ^1H spin diffusion. As can be seen in Table 1, the two noncrystalline lines are associated with well-distinguishable T_{1H} values. That is, distinctly different T_{1H}'s, 1.61 and 0.39 s, can be recognized. The former value indicates a restricted molecular motion in the T_{1H} time frame. Table 1 also contains the data of another sample with molecular weight of 248,000. For this sample, the T_{1H} of the crystalline phase was measured to be 2.20 s. We note that this value is very close to 2.04 s for the line at 31.3 ppm, whereas it is significantly longer than 0.50 s for the line at 31.0 ppm. This shows that the noncrystalline

Table 1. Linear polyethylene samples and their characterization

Sample				Crystalline components		Noncrystalline components	
Molecular weight $\overline{M_v}$	Crystallinity $(1-\lambda)_d^a$	$(1-\lambda)_{BL}^b$	Parameter	Monoclinic (34.4 ppm)	Orthorhombic (33.0 ppm)	Interfacial (31.3 ppm)	Rubbery (31.0 ppm)
Bulk crystals							
3×10^6		0.620	mass fraction	≈ 0	0.66	0.18	0.16
			half width (Hz)		18	85	38
			T_{1H} (s)		1.87	1.61	0.39
			T_{1C} (s)		2560, 263, 1.7	0.37	0.37
			T_{2C} (ms)			0.044	2.4
248000	0.761	0.755	mass fraction	0.06	0.70	0.16	0.08
			half width (Hz)	50	18	86	37
			T_{1H} (s)		2.20	2.04	0.50
			T_{1C} (s)		2750, 111, 1.3	0.41	0.41
Solution crystal							
91000	0.834	0.780	mass fraction	0.09	0.68	0.23	≈ 0
			half width (Hz)	60	20	133	
			T_{1H} (s)		1.90	1.90	
			T_{1C} (s)		220, 21, 2.0	0.46	
			T_{2C} (ms)			0.04	

[a] Crystallinity from density measurements. [b] Crystallinity from broad-line ^1H NMR.

component, which comprises the crystalline-amorphous interphase, and whose contribution to the spectrum appears at 31.3 ppm, is located adjacent to the crystalline component so that ^{1}H spin diffusion occurs between these two phases. In Table 1 the data of the sample crystallized from dilute solution are also listed. We will discuss these data in the following section.

Molecular Weight Dependence of Phase Structure. Similar line shape analysis was performed for samples with molecular weight over a very wide range that had been crystallized from the melt. In some samples, an additional crystalline line appears at 34.4 ppm which can be assigned to *trans-trans* methylene sequences in a monoclinic crystal form. Therefore the spectrum was analyzed in terms of four Lorentzian functions with different peak positions and line widths; i.e. for two crystalline and two noncrystalline lines. Reasonable curve fitting was also obtained in these cases. The results are plotted by solid circles on the data of the broad-line ^{1}H NMR in Fig. 3. The mass fractions of the crystalline, amorphous phases and the crystalline-amorphous interphase are in good accord with those of the broad, narrow, and intermediate components from the broad-line NMR analysis.

Provided the crystalline stem length ζ_c is known and a stacked lamellar structure is assumed, the thicknesses of the interphase and amorphous phase can be evaluated from these data by the Eqs. (6) and (7):

$$\zeta_i = \zeta_c x_i / 2x_c \qquad (6)$$

$$\zeta_a = \zeta_c x_a / x_c \qquad (7)$$

where ζ_i and ζ_a are the thicknesses of the interphase and amorphous phase, and x_c, x_a, and x_i designate the mass fraction of the crystalline, amorphous, and interphases, respectively. For the evaluation of ζ_i and ζ_a, for samples with different molecular weights, we used the values for the crystalline stem length ζ_c that were reported by Voigt-Martin and Mandelkern [56], and by Bassett et al. [57] for the linear polyethylene samples crystallized under the same condition. The result is shown in Fig. 10, where the thickness of each phase is plotted against molecular weight over a very wide range. Samples with very low molecular weights consist only of a lamellar crystalline phase and a crystalline-amorphous interphase (in this molecular weight range the interphase may be considered as a noncrystalline overlayer, since the amorphous rubbery phase is not present). The thickness of the interphase gradually increases and levels off as the molecular weight exceeds 30,000 with the appearance of the rubbery phase. The thickness of the interphase stays unchanged at 34Å in the range 30,000~100,000. It increases again with enhanced increasing of the thickness of the rubbery phase as the molecular weight further increases above 100,000.

The fact that the thickness of the interphase estimated here stays unchanged at 34Å in the molecular weight range of 30,000~100,000, while the mass fraction and thickness of amorphous phase change remarkably, is particularly meaningful. Flory et al. [6, 7] anticipated in 1984 based on their lattice theory that the methylene chains that emerge from the basal plane of lamellar crys-

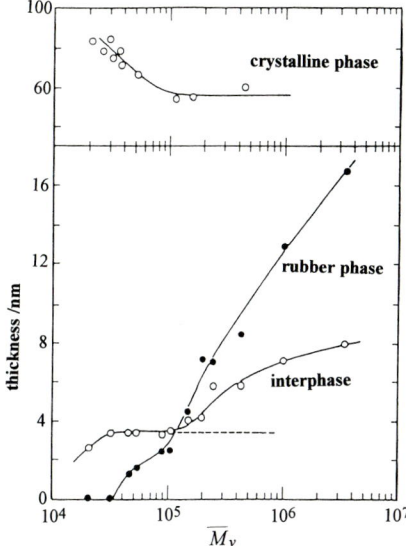

Fig. 10. Thickness of the crystalline, amorphous, and crystalline-amorphous interfacial phases of bulk polyethylene as a function of molecular weight $\bar{M}_v$

tallites must traverse the crystalline-amorphous interphase as a transition region of a thickness of ca. 25Å to reach the amorphous phase where methylene chains have no correlation with the crystalline phase. This thickness of 34Å is not so far from the theoretical expectation of 25Å, and since the latter value can be changed by parameters assumed in their calculation, the result obtained here confirms experimentally their theoretical expectation. As the molecular weight increases above 100,000 to 3×10^6, the thickness of the interphase increases and reaches 80Å. This is because the stacked lamellar structure assumed is no longer adequate for samples with such large molecular weights. The samples studied here were obtained by the isothermal crystallization at 130 °C, followed by slow cooling. In samples with very large molecular weights, additional crystallization may occur during the cooling and the structure will become more complicated so that the simple stacked lamellar crystalline structure cannot be assumed.

3.3
Discussion

It is evident that the noncrystalline component is distributed in two phases that are associated with the same T_{1C} but different T_{2C} values. What does this mean? In order to understand this phenomenon we have to refer to the theory of the relaxation reviewed in Section 2.2 [43]. Provided the internuclear vector between carbon and hydrogen nuclei involves only a single motion, that is, if each term of the correlation function of the dipole-dipole interaction between ^{13}C and ^1H spins evolves exponentially with one correlation time τ_C (relaxation time of motion) according to Eq. (3), the T_{2C} must be same when T_{1C} is the same. This fact cannot be understood if these two phases are not assumed to involve more than two independent motions dictated by different τ_C values.

Consider the relation between the correlation function and its spectral density. Slower and faster decays of the correlation function (i.e. slower and faster motions) give narrower and wider distributions of the spectral density, respectively. Figure 11 (a) shows some decay curves of the correlation function for motions with different τ_C values and (b) indicates the distributions of their respective spectral densities that are obtained by Fourier transform of the decay curves. Here A, B, C in (a) are the decay curves with τ_{CA}, τ_{CB}, τ_{CC} and A, B, C in (b) are the distribution of their respective spectral densities. As can be seen, the decay becomes slower and the spectral density distribution becomes narrower as the τ_C increases. Assume here that $\tau_{CA} \ll \tau_{CB} \ll \tau_{CC}$ and the amorphous phase involves two independent motions dictated by τ_{CA} and τ_{CB} whereas the crystalline-amorphous interphase involves two motions dictated by τ_{CA} and τ_{CC}. Here, τ_{CA} characterizes a local molecular motion with relation to few carbon atoms in the main molecular chain, and τ_{CB}, τ_{CC} a somewhat long-ranged motion with relation to a conformational change of ca. 10–20 carbon atoms. In other words, it is assumed that somewhat long-ranged motion is different between the two phases but local motion is the same, the former is dictated by τ_{CB} or τ_{CC} and the latter by a common relaxation time τ_{CA}.

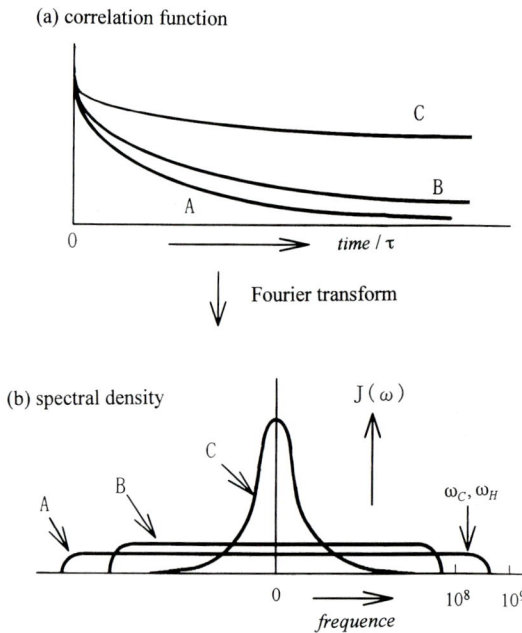

Fig. 11. a The free induction decay (correlation function) with different relaxation times τ_c's, and **b** Fourier transforms (spectral density)

As can be seen from Eq. (1), T_{1C} is determined by the spectral densities at the frequencies of ω_C, and $\omega_H \pm \omega_C$. Assume here that the distributions of the spectral densities are as shown in Fig. 11–(b), where the frequency range of ω_H, ω_C, and $\omega_H \pm \omega_C$ is shown by the vertical arrow. Then the T_{1C}'s of both of the noncrystalline phases are determined by the local motion dictated by τ_{CA}, independent of the other motions that have no meaningful spectral density at the frequency range indicated by the arrow. Then the two noncrystalline phases must have the same T_{1C}. On the other hand, the T_{2C} is determined by the spectral density at zero for the solid matter as pointed out at the end of Sect. 2.2. Hence, the T_{2C}'s of the amorphous phase and the crystalline-amorphous interphase are respectively determined by the motion of B and C, because the spectral density at zero of motion A is negligible. Since the spectral densities at zero frequency of both the B and C motions are quite different, the two phases must have quite different T_{2C} values, [58].

To this end, we emphasize that the two phases assumed for the noncrystalline component are well-defined by the differences in T_{2C}. In the crystalline-amorphous interphase, a long-ranged molecular motion accompanying the conformational change of ca. 10–20 carbon atoms is very slow or almost inhibited. In this phase the *trans-trans* conformation may be somewhat abundant but generally all permitted conformations are widely distributed.

4
Linear Polyethylene Crystallized from Dilute Solution

4.1
Introduction and Approach by Broad-Line ^{1}H NMR

When linear polyethylene is crystallized from dilute solution, lozenge-shaped crystallites are observed under an electron microscope that are sometimes called single crystals [59–61]. The thickness of the lozenge-shaped crystallites is generally smaller than 10 nm, whereas the width exceeds several µm, and the molecular chains are oriented approximately perpendicular to the wide lozenge faces. The problem here is the molecular chain conformation in the boundary between the crystalline lamellae or on the lamellar crystallite surface. Since the crystal stem length is much shorter than the extended molecular chain length for samples with an ordinary molecular weight, the molecular chains have to participate in such a lozenge-shaped crystallite repeatedly by regular or irregular folding.

Our broad-line ^{1}H NMR analysis showed that this type of sample generally consists of the phase structure of lamellar crystallites and noncrystalline overlayer with a negligible amount of the noncrystalline amorphous phase [16, 62]. In broad-line ^{1}H NMR spectra of solution-grown linear polyethylene samples, a narrow component that suggests the existence of a liquid-like amorphous phase is hardly recognized. In Table 2, the three-component analysis of the broad-line ^{1}H NMR spectra of linear polyethylene samples with different molecular weights that were crystallized isothermally from 0.08% toluene solution at 85 °C for 24 hours under a nitrogen atmosphere is summarized.

Table 2. Three-component analysis of the broad-line ^{1}H NMR spectra of solution-grown linear polyethylene samples with different molecular weights

	Mass fraction		
$\overline{M_\eta}$	w_b	w_m	w_n
17600	0.861	0.136	0.003
106000	0.825	0.169	0.006
3400000	0.835	0.160	0.005

w_b, w_m, and w_n indicate the massfraction of broad, medium, and narrow components, respectively.

The mass fraction of the narrow component that corresponds to the rubbery noncrystalline amorphous phase is as small as 0.003–0.006. The mass fraction does not increase appreciably with increasing temperature, but stays almost unchanged up to 70 °C. Hence, it is concluded that solution-grown samples do not actually comprise a rubbery amorphous phase. This conclusion is confirmed by high-resolution solid-state ^{13}C NMR with more detailed information.

4.2
Approach by High-Resolution ^{13}C NMR

Figure 12 shows the DD/MAS ^{13}C equilibrium spectrum of a solution-grown polyethylene with a viscosity-average molecular weight of 91,000 at room temperature. In order to analyze this spectrum, and determine the content of the structural components of this sample, the spin-lattice and spin-spin relaxation phenomena were examined by similar techniques to those employed for the study of the bulk-crystallized samples. Firstly, the spin-lattice relaxation that suggests the existence of two different noncrystalline components was not seen. The T_{1C} was estimated to be 0.46 s. Hence, the noncrystalline component was judged to comprise a uniform phase as far as the T_{1C} value was concerned. To examine further the content of this noncrystalline component, the spin-spin relaxation was examined. That is, the partially recovered magnetization in the z direction for τ_ℓ of 3.5s was relaxed in the transverse direction (i.e. in the xy plane) for a varying time τ_t and the FID was observed under ^{1}H DD. In the spin-spin relaxation pulse sequence, all noncrystalline magnetizations quickly disappeared simultaneously within τ_t = 100 μs. The T_{2C} was estimated to be ca. 40μs. Therefore, the noncrystalline component of the solution-grown sample evidently forms a uniphase in as much as judged by both T_{1C} and T_{2C}. The T_{1C} of 0.46 s and T_{2C} of ca. 40μs are almost equivalent to those of the crystalline-amorphous interphase of the bulk-crystallized polyethylene (cf. Table 1). This suggests that the molecular mobility and conformation of the noncrystalline component of the solution-grown sample are similar to those in the crystalline-amorphous interphase of the bulk-crystallized samples.

We have examined solution-growth polyethylene samples which differ in molecular weight over a wide range. However, it was found that this characteristic phase structure does not change appreciably in relation to the molecular weight.

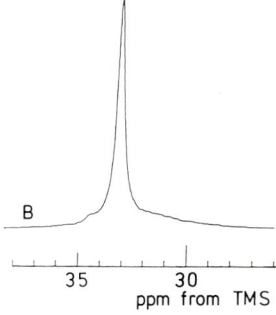

Fig. 12. 50 MHz DD/MAS ^{13}C NMR spectrum of solution-grown polyethylene with a viscosity-average molecular weight of 91,000 at room temperature. The spectrum was obtained by pulse sequence I (shown in Fig. 1) with the repetition time, $\tau_\ell = 1{,}500$ s

4.3
Discussion

The lozenge-shaped crystallites of solution-grown samples are sometimes assumed to be formed by regularly folding molecular chains. The neighboring crystalline stems are assumed to be connected tightly by a regularly folded methylene sequence such as a *ggtgg* conformation. However, the NMR results cited above do not allow such regular folding molecular chains to form the lozenge-shaped crystallite. If the crystal methylene sequence of the trans *zig-zag* conformation is connected by a *ggtgg* sequence, additional resonant lines assignable to the *gt-tt*, *gg-tt*, *tg-gt*, *gt-gg* methylene carbons are expected in the ^{13}C NMR spectrum upfield to the resonance line assignable to the crystalline trans-trans methylene sequence, i.e. in the range 30–22 ppm. Here the *gt-tt*, for example, designates the methylene carbon both of whose sides are adjacent to *gt* and *tt* methylene sequences, respectively. In fact, such resonance lines were distinctly recognized for crystalline cycloalkanes [c–$(CH_2)_n$] [63, 64]. I. Ando et al. have also reported the appearance of such resonance lines in their CP/MAS spectrum of solution-grown polyethylene [65]. We have examined critically the DD/MAS ^{13}C NMR spectrum of a polyethylene sample that was crystallized isothermally from dilute solution under strictly controlled conditions. However, we could not find any trace of such resonance lines. The resonance lines of the regularly folded methylene sequence become vague with increasing temperature above room temperature because of the broadening of each line even for well-crystallized c–$(CH_2)_{60}$.

For this cycloalkane, the T_{1C}'s of the methylene carbons in the folded sequence are estimated to be 35–39 s at room temperature and become shorter with increasing temperature [66]. Hence, the T_{1C}'s for the regularly folded methylene sequence in polyethylene, if it exists, will be appreciably shorter than 35 s at room temperature, and much shorter than ca. 220 s of the inner methylene carbons of polyethylene (cf. Table 1). Hence, if we measure the DD/CP ^{13}C NMR spectrum of the solution-grown polyethylene sample by a single pulse sequence with a repetition time of ca. 50 s, the contribution from the folded methylene sequence should be recognizable because of the reducing intensity of the inner methylene carbons. However, we could not find any trace of such resonance lines. Hence, we have to suppose that I. Ando et al. took into account some noise in their CP/MAS ^{13}C NMR spectrum, because of insufficient signal/noise ratio, as the resonances from the folded methylene sequences, and that their experimental results must be revised. Thus, we conclude that the lozenge-shaped single crystal of polyethylene is not formed by connecting the crystal sequence with the regularly folded methylene sequence.

5
High-Pressure Crystallized Polyethylene

5.1
Introduction

It is well-known that when linear polyethylene is crystallized from the melt at high pressure and high temperature, a structure of extremely thick crystal lamellae is produced, sometimes called an extended crystal. In the preceding sections, we examined the phase structure of linear polyethylene, crystallized either from the melt or dilute solution at atmospheric pressure. However, the phase structure of linear polyethylene is thought to differ widely according to its crystallization conditions, thermal and mechanical history, if the material is the same. In this section, we examine samples crystallized under high pressure with the combined use of NMR and electron microscopy [18]. The materials chosen were crystallized at high pressure, to produce samples of the highest crystallinity which are arguably the least complicated of melt-crystallized polyethylene, so enhancing the prospects of obtaining definitive conclusions.

Table 3. Characterization of pressure-crystallized linear polyethylene samples

Sample	M_w[a]	M_w/M_n	Crystallinity		Crystal thickness(nm)	T_m[d](°C)
			Density[b]	^{1}H NMR[c]		
Fraction 2	28600	1.22	0.971	0.928	457[e]	140.0
Fraction 1	33000	1.25	0.972	0.944	580[e]	140.0
Rigidex 9	132000	10.4	0.928	0.898	82[f]	140.3
HO20-54P	231000	8.2	0.931	0.878	120[f]	142.3
Hizex 1900	1.1×10^6	–	0.869	0.834	190[f]	147.2

[a] Estimated by g.p.c.
[b] Estimated from density measurements.
[c] Obtained by broad-line ^{1}H NMR.
[d] Peak temperature in a DSC thermogram measured at a rate of 10 K/min.
[e] Number-average crystal thicknesses determined by electron micrography.
[f] Number-average crystal thicknesses determined by nitrogen/g.p.c. method.

5.2
Experimental

Six linear polyethylene samples were examined and the results are collected in Table 3. They were melted at 260 °C for 1 h at 4.95 kbar and then cooled slowly. Cooling rates over the range of the crystallization temperatures were 0.25 K/min for two molecular weight fractions with lower molecular weights (Fraction 1 and 2) and 2 K/min for the other samples. As can be seen, these high-pressure crystallized samples are generally characterized by very thick lamellae with high-melting temperatures. However, their crystal thickness is still much shorter than their extended molecular lengths except for the two samples with lower molecular weight. Hence, the lamellar structures of these samples are thought to be formed by folding molecular chains in essentially the similar manner as samples crystallized at atmospheric pressure.

Broad-line ^{1}H NMR. The line shape analysis of broad-line ^{1}H NMR spectra into the broad, medium, and narrow components was carried out for these samples, and they were assigned to the crystalline, interfacial, and rubbery amorphous phases, respectively. Their mass fractions are shown in Fig. 13 as a function of weight-average molecular weight. In Fig. 13, the data of samples crystallized at atmospheric pressure shown in Fig. 2 are also shown for comparison by a broken line. When crystallized at atmospheric pressure, samples consist only of the crystalline phase and a noncrystalline overlayer in the molecular weight range lower than 20,000 as pointed out previously. Above this the amorphous phase with rubbery molecular motion appears and its mass fraction continues to increase with molecular weight; the noncrystalline overlayer also starts to increase above a molecular weight of 20,000. On the other hand, for the high-pressure crystallized samples, the rubbery amorphous phase does not appear below a molecular weight of 100,000, and is only just discernible in Rigidex 9 with a molecular weight of 132,000 (the mass fraction of the amorphous phase can be

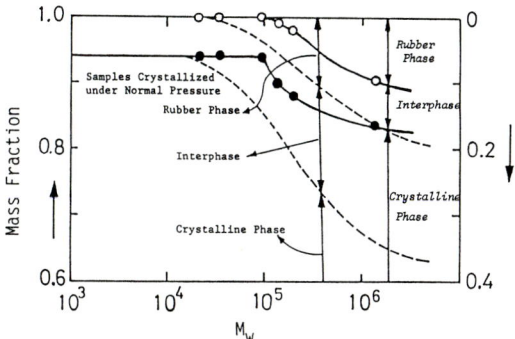

Fig. 13. Mass fractions of three phases of bulk polyethylene versus weight-average molecular weight, obtained by broad-line ^{1}H NMR analyses. *Solid lines* indicate high-pressure crystallized samples, while *broken lines* show samples which were isothermally crystallized under atmospheric pressure

read by the upper solid line with data shown by open circles as the distance from the upper horizontal coordinate axis). Beyond this molecular weight, the rubbery amorphous phase increases concomitantly with decreasing crystallinity but to a lesser extent in comparison with that found for the samples crystallized at atmospheric pressure; even above a molecular weight of 10^6 its mass fraction hardly exceeds 0.10, and the crystallinity remains as high as 0.80. This seems to be characteristic of the high-pressure crystallized samples.

High-Resolution ^{13}C NMR Spectrum. Figure 14 shows the DD/MAS ^{13}C NMR spectra of pressure-crystallized samples of Fraction 1, Rigidex 9, HO20-54P, and Hifax at room temperature. For all samples, there are distinctly recognizable sharp peaks assignable to the orthorhombic crystalline component at ca. 33 ppm (Peak I) and the noncrystalline component at ca. 31 ppm (Peak II). For samples HO20-54P and Hifax, the resonances assignable to a monoclinic crystalline and noncrystalline components are further recognized at ca. 34.4 ppm (Peak III) and 31 ppm, respectively. For other samples, the existence of the noncrystalline contribution is a little vague due to the high crystallinity, but the existence was clearly confirmed by enhancing the noncrystalline contribution using a shorter τ_ℓ in the pulse sequence $(90°-\text{FID}_{DD}- \tau_\ell)_n$ for obtaining ^{13}C NMR spectrum.

Spin-Lattice and Spin-Spin Relaxation. In order to examine the content of these crystalline and noncrystalline components, we examined the spin relaxation of each resonance line. Firstly it was found that the line due to the orthorhombic crystalline component at 33 ppm involves plural T_{1C}'s for all samples, as summarized in Table 4. In relation to the T_{1C} for each sample, very long T_{1C} values are recognized for three higher molecular weight samples. These values are expected

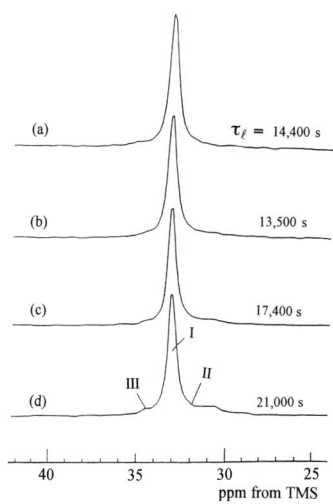

Fig. 14. DD/MAS ^{13}C NMR spectra of different polyethylene samples crystallized under high pressure, obtained by a π/4 single pulse sequence [π/4-FID-τ_ℓ]; (a) Fraction 1, (b) Rigidex 9, (c) HO20-54P, (d) Hifax. The repetition time τ_ℓ is indicated in each spectrum

from their very thick crystal lamellae as extrapolated values from the values reported for samples with thinner lamellar thicknesses [17, 67]. The T_{1C}'s of two fractionated samples with lower molecular weights are also longer than 4,000 s. However, these T_{1C}'s are still short, if their very thick crystal lamellae are considered. The crystal thickness of these samples is equivalent to or longer than the extended molecular chain length. Hence the lamellar structure of these samples is thought to be formed mostly by fully extended molecular chains without chain folding so that some methyl molecular chain ends may be incorporated inside the crystal lamellae. This will cause the relatively shorter T_{1C} against the very thick crystal lamellae. For all samples there are recognized plural T_{1C} values in a very wide range of several thousands, several hundreds, several tens, and a few seconds as can be seen in Table 4. These plural T_{1C} values seem to correspond to the wide distribution of the lamellar thicknesses observed by electron microscope.

On the other hand, the T_{1C} of the noncrystalline resonance at ca. 31 ppm is found to be in the range of 0.35–0.50 s for all samples. When judged only by the T_{1C}, the noncrystalline component forms a monophase for all samples. However, the noncrystalline line involves distinctly different T_{2C} values at least for the three larger molecular weight samples. As an example, ^{13}C spin-spin relaxation behavior of the noncrystalline line for Hifax is shown in Fig. 15. The spin-spin relaxation behavior was examined by the pulse sequence $(90°–\tau_t –FID_{DD}– \tau_\ell)_n$, with $\tau_\ell = 10s$ and varying τ_t. The height of the peak of the noncrystalline component at 31 ppm in the course of this spin-spin relaxation is plotted against τ_t. The overall decay curve (the upper line in Fig. 15) can be clearly resolved into two parts, a rapid decay and a subsequent slow decay. The slope of the slow decay yielded $T_{2C} = 2.02$ ms, and the T_{2C} of the rapid decay was estimated to be 0.020 ms by the usual decay analysis as shown at the bottom of Fig. 15. This indicates that the noncrystalline component comprises two different phases, both associated with the same T_{1C} of 0.35 s and distinctly different T_{2C}'s of 2.02 ms and 20 μs. The former is assigned to the amorphous phase with the rubbery molecular motion and the latter to the crystalline-amorphous interphase. The T_{1C} and T_{2C} values for all samples are summarized in Table 4.

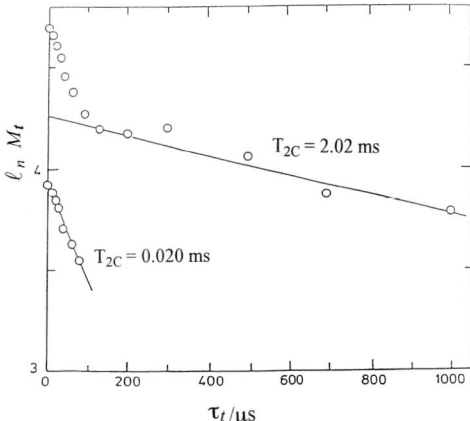

Fig. 15. ^{13}C spin-spin relaxation behavior of the noncrystalline component of Hifax

Table 4. ^{13}C spin-lattice and spin-spin relaxation times of high-pressure crystallized polyethylene

Sample	T_{1C} (s)		T_{2C} (ms)	
	Crystalline (33 ppm)	Noncrystalline (31ppm)[b]	Interfacial[c]	Amorphous
Fraction 2	4510, 526, 12	0.45	0.049	a
Fraction 1	4720, 683, 11	0.50	0.050	a
Rigidex 9	4320, 686, 46, 2	0.45	0.061	0.86
HO20-84P	5710, 656, 21	0.40	0.010	2.2
Hifax 1900	7010, 810, 20, 2	0.35	0.020	2.0

[a] The mass fraction is negligible so that T_{2C} is not measurable.
[b] Both of the amorphous and crystalline-amorphous interphase are associated with the same T_{1C}.
[c] For samples without an amorphous component the interfacial material may be considered as a noncrystalline interlamellar material.

As can be seen from Table 4, two lower molecular weight samples actually comprise only the crystalline and noncrystalline interlamellar material, devoid of amorphous phase. On the other hand, the larger molecular weight samples comprise three phases; the crystalline, amorphous, and crystalline-amorphous interphase in a similar fashion to the atmospheric-pressure crystallized samples. However, we note that the T_{2C}'s of the crystalline-amorphous interphase for two higher molecular weight samples is appreciably shorter than those of the atmospheric-pressure crystallized samples. This demonstrates that the molecular chain motion in the crystalline-amorphous interphase of these pressure-crystallized samples on a T_{2C} time scale is more severely restricted.

Lineshape Analysis and Phase Structure. Since the pressure-crystallized samples generally comprise a crystalline and two noncrystalline phases, we analyzed the equilibrium spectra of two higher molecular weight samples shown in Fig. 14, and the results are summarized in Table 5. If we assume the typical stacked lamellar structure for these samples, the thickness of each phase can be estimated by using the mass fraction obtained from the line shape analysis and the crystalline stem length obtained by other techniques. The thicknesses of the crystalline-amorphous interphase and the amorphous phase, ζ_i and ζ_a, respectively, may be estimated by Eqs. (6) and (7) with $x_c = x_{c1} + x_{c2}$ where x_{c1} and x_{c2} denote the volume fractions of the orthorhombic and monoclinic crystalline phases. Since the values of the crystalline stem length or the lamellar thickness involve a wide distribution, and also depend on the method of estimation, the situation is not so simple. Here, however, the weight-average crystal thickness, estimated from the nitric acid degradation/g.p.c. technique [68], was used. The results are shown in Table 5, together with the mass fractions of each phase.

As can be seen, for both samples the crystalline mass fraction (orthorhombic plus monoclinic) agrees well with that estimated from the density measurements and broad-line ^1H NMR analysis. The thicknesses of the interphases are 3.8 and

Table 5. Mass fraction and thicknesses of crystalline and noncrystalline phases

Sample	Mass fraction				Thickness of each phase* (nm)		
	Crystalline		Noncrystalline				
	Orthorhombic	Monoclinic	Interfacial	Amorphous	Crystalline	Interfacial	Amorphous
HO20-54	0.874	0.017	0.055	0.053	120	3.8	7.3
Hifax	0.843	0.037	0.071	0.049	90	8.0	11.0

* Estimated from the crystalline stem lengths (or lamellar thicknesses) and mass fractions of the phases (refer to the text). Here the volume fraction of all phase was assumed to be equal to the mass fraction, i.e. the densities of all phases were assumed to be unity. This assumption may not introduce any appreciable influence on the discussion of this work.

8.0 nm for HO20-54P and Hifax, respectively. The value obtained for HO20-54P is almost the same as the 3.4 nm reported for the atmospheric-pressure crystallized samples with normal molecular weights in the range 30,000–150,000 (cf. Fig. 10). On the other hand, the value for Hifax is significantly larger than the values cited in the preceding sections for other types of samples as well as the values of 1.5–2.0 nm which were evaluated theoretically by Flory et al. [7]. An increase in the thickness of the interfacial phase was also noted for the atmospheric-pressure crystallized samples if the molecular weight exceeded 150,000, as mentioned previously.

5.3
Discussion

As pointed out previously with relation to Fig. 13, the dependence of the phase structure on molecular weight shifts to the higher molecular weight side in comparison with samples crystallized at atmospheric pressure. In particular, the rubbery amorphous component appears above a molecular weight of 30,000 for atmospheric-pressure crystallized samples, but above a value of 100,000 for high-pressure crystallized samples. The significance of this appears to lie in the polydispersity of the samples. The atmospheric-pressure crystallized samples examined in the previous section were fractions with a molecular weight polydipersity ($\overline{M_w}/\overline{M_n}$) of ca. 2, so at $\overline{M_w}$ = 30,000 the number-averaged extended molecular chain length X_n is approximately twice the number-averaged crystal thickness ζ_n. Below the limit of 2 for the ratio X_n/ζ_n, a molecular chain participates only once in a crystallite and so-called extended crystals are produced. If the ratio X_n/ζ_n exceeds 2, a molecular chain participates repeatedly in a crystallite by chain folding and the amorphous phase is thought to appear. On the other hand, with regard to the pressure-crystallized samples, the rubbery phase could only just be detected by the broad-line ^{1}H NMR analysis even for the Rigidex 9 sample with $\overline{M_w}$ = 132,000 ($\overline{M_n}$ = 13,200). The fit here does not at first sight appear so good, since the number-averaged chain length (115 nm) is not much bigger than the number-averaged crystal thickness (82 nm). However, during high-pressure crystallization considerable molecular fractionation into discrete crystal populations occurs, and taking this into account there is good reason to suppose that the same presence or absence of repeat folding is the determining factor. For the samples with higher molecular weights that involve the rubbery amorphous component, X_n/ζ_n and X_w/ζ_w are ~ 2.2 and 7.5–26, respectively. It is then reasonable to assume, even in the pressure-crystallized samples, that a molecular chain participates repeatedly in a crystallite or several crystallites by chain folding and the amorphous phase appears.

6
Ultra High Modulus Polyethylene

6.1
Introduction

Recently, various techniques that produce highly oriented linear polyethylene with a ultra high modulus (hereafter, referred to as UHMPE) have been developed. In this section, we will examine the structure of the UHMPE that was prepared by highly drawing a dried gel [69]. Even if bulk polyethylene is uniaxially highly drawn by a normal method at a temperature between the T_g and T_m, the phase structure is essentially similar to the undrawn sample. That is, it involves three phases of the crystalline and two noncrystalline phases, although the mass fraction and detailed content of each phase are somewhat different. However, UHMPE samples may have a particular phase structure.

6.2
Experimental

The first samples examined were prepared by the method developed by Smith et al. [70, 71] and by Matsuo [72]. Sample films of a thickness of ca. 100 μm were obtained by drying a gel which was obtained by quenching a 0.4 g/dl decalin solution of linear polyethylene with a molecular weight of 3×10^6 from 140 °C in icewater. The samples thus obtained could be drawn to a very high extent because of very few intermolecular chain entanglements. However, since they could not be drawn highly in one step, they were drawn 10 times at the first step in a silicon oil bath at 145 °C at a rate of 1.6 times/min and then at the second step they were drawn so that the final draw ratio was 50, 100, and 150 times.

DD/MAS ^{13}C NMR Spectra. Figure 16 shows the equilibrium ^{13}C NMR spectra of those samples that were obtained by the pulse sequence $(\pi/4-\text{FID}_{DD}-3T_{1C})_n$ at room temperature. Here the sample A is the undrawn dried gel, B is the sample obtained from A by annealing at 145 °C for 4 minutes, and C, D, and E are samples drawn 50, 100, and 50 times, respectively. For most spectra there is a recognized downfield resonance at ca. 33 ppm assignable to the orthorhombic crystalline component and an upfield resonance at ca. 31 ppm assignable to the noncrystalline component.

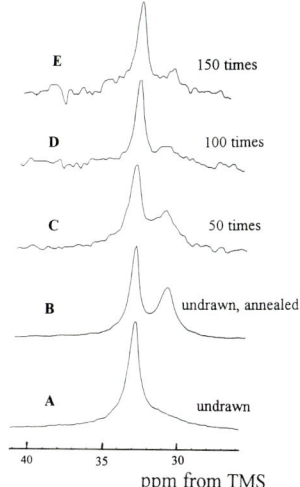

Fig. 16. 50 MHz DD/MAS ^{13}C NMR spectra of highly drawn polyethylene samples with different draw ratios, obtained by a π/4 single pulse sequence (π/4-FID-3T$_{1C}$)

Table 6. ^{13}C Spin relaxation times of ultra-drawn polyethylene films

Sample	Draw ratio	T_{1C}/s				T_{2C}/ms	
		Crystalline			Noncrystalline	Interfacial	Amorphous
A[a]	1	55	7.1	1.8	0.46	0.023	–
B[b]	1	278	18.2	2.0	0.45	0.054	5.5
C	50	599	21.1	2.3	0.47	0.060	4.4
D	100	1018	86.4	2.0	0.56	0.032	2.8
E	150	1650	134.4	1.5	0.57	0.078	7.7

[a] Dried gel sample. [b] Sample obtained by annealing sample A at 145 °C for 4 min.

Spin-Lattice and Spin-Spin Relaxations. In order to determine the content of these crystalline and noncrystalline resonances, the longitudinal and transverse relaxations were examined in detail. It was first confirmed that the noncrystalline resonance of all samples is associated with T_{1C} in an order of 0.45–0.57 s. Hence, the noncrystalline component of all samples comprises a monophase, in as much as judged only by T_{1C}. However, it was found that the noncrystalline component of drawn samples generally comprises two phases with different T_{2C} values; amorphous and crystalline-amorphous interphases. The dried gel sample does not include rubbery amorphous material; it comprises the crystalline and rigid noncrystalline components. However, the rubbery amorphous phase with T_{2C} of 5.5 ms appears by annealing at 145 °C for 4 minutes. For the orthorhombic crystalline component, three different T_{1C} values, that suggest the distribution of crystallite size, were recognized for each sample, as normal for crystalline polymers [17, 54, 55]. The T_{1C} and T_{2C} of all samples examined are summerized in Table 6.

Table 7. ^{13}C Chemical shift and mass fraction of each phase of ultra-drawn polyethylene

Sample	Draw ratio	Chemical shift/ppm				Mass fraction			
		Ortho-rhombic	Mono-clinic	Inter-face	Amor-phous	Ortho-rhombic	Mono-clinic	Inter-face	Amor-phous
A[a]	1	32.9	34.7	31.2	—	0.545	0.159	0.297	0
B[b]	1	32.9	34.7	31.6	30.7	0.442	0.041	0.127	0.39
C	50	32.9	34.8	31.8	30.8	0.451	0.089	0.181	0.279
D	100	32.9	34.8	31.7	30.8	0.682	0.109	0.078	0.130
E	150	32.9	34.8	31.7	30.7	0.688	0.135	0.053	0.134

[a] Dried gel sample.
[b] Sample obtained by annealing sample A at 145 °C for 4 min.

The longest T_{1C} of each sample increase with the increases in draw ration. However, even for 150 fold drawn sample (E) the T_{1C} does not exceed 1650 s. This seems to reflect the phase structure of this type of sample. In addition to the orthorhombic crystalline resonance at 32.9 ppm, the existence of downfield resonance at 34.7–34.8 ppm assignable to the monoclinic crystalline component for all samples was confirmed by examining the relaxation phenomena.

6.3
Phase Structure and Discussion

The mass fractions of all phases of each sample were obtained by line shape analysis of the equilibrium spectra, and the results are summarized in Table 7.

As can be seen, the crystalline fraction (orthorhombic plus monoclinic) decreases from 0.70 to 0.54 at a draw ratio of 50 times and increases to 0.82 at the largest draw ratio. The amorphous phase that appears at a lower draw ratio decreases with increasing draw ratio, accompanying the decrease of the interphase. Such phase structure as elucidated here will reflect on the various macroscopic properties of samples.

Figure 17 shows the thermograms of these samples measured with a differential scanning calorimeter at a rate of 10 °C/min. The temperature at which melting is complete is extraordinarily high for most samples with the exception of the annealed undrawn sample (sample B) in comparison to the reported values for linear polyethylene. The melting temperature of sample A (undrawn dried gel sample) is 147.0 °C, which exceeds the equilibrium melting temperature that was obtained for this polymer by theoretical extrapolation from the melting temperatures of n-paraffins by Flory et al. [73] or by the dependence of the observed melting temperatures on the crystallization temperature by Mandelkern et al. [74]. Hence, the reason for such a high melting temperature could be attributed to the very high viscosity at the molten state of this sample due to the very high molecular weight. That is, complete randomization of molecular chains does not stop when melting is complete while heating at a rate of 10 °C/min due to the very high viscosity at the melt. Hence, the observed high melting temperature can be attributed to the decrease of entropy change of the fusion.

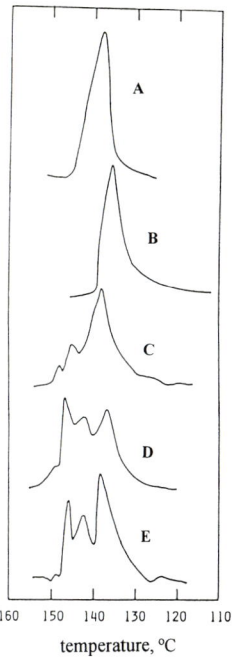

Fig. 17. DSC thermograms of highly drawn polyethylene samples with different draw ratios. *A*: undrawn, *B*: undrawn, annealed, *C*: 50 times drawn, *D*: 100 times drawn, *E*: 150 times drawn

As mentioned above, 0.39 fraction of the amorphous phase appears upon annealing at 145 °C for 4 minutes for this sample (see Table 7). In compliance with this change, the melting temperature of sample B decreases to 138 °C, which is slightly higher than the melting temperatures observed for well-crystallized samples of this polymer. However, upon drawing, the melting temperature increases extraordinarily and three well-distinguishable endothermal peaks are observed. For example, for the 50 fold drawn sample (sample C) the melting temperature increases to over 150 °C and two small additional endothermal peaks at 145.0 and 147.9 °C are observed in addition to the main peak at 137.9 °C before complete fusion. Three similar endothermal peaks in the fusion are reported for a fibrous crystal of ultra-high molecular weight polyethylene [75], a uniaxially drawn sample [76] and a drawn sample of single crystal mat [77]. These endothermal peaks were attributed from the low temperature side to be the melting of randomly oriented crystallites, the solid-solid phase transformation from orthorhombic to hexagonal crystal form, and the melting of the hexagonal crystal [69, 70]. However, since the melting of the hexagonal crystal could not be defined for a highly drawn product of a single crystal mat, Furuhata et al. pointed out that the intermediate peak could not be attributed to the solid-solid crystal transformation [71]. We note here that the intensity of the highest temperature peak increases remarkably upon drawing (cf. the isotherm for 100 and 150 fold drawn samples). Such an intensity increase cannot be understood by the rea-

Table 8. Phase structure of ultra-high modulus polyethylene obtained by solid-state coextrusion and drawing at 135 °C

Sample	Draw ratio	Chemical shift/ppm			Mass fraction			
		Ortho-rhombic	Mono-clinic	Inter-face	Ortho-rhombic	Mono-clinic	Inter-face	Amor-phous
PUMG	1	32.9	–	31.2	0.808	0	0.192	0
PUM5	5	32.9	34.1	31.0	0.818	0.035	0.147	0
PUM5A	5	32.9	34.1	31.4	0.861	0.042	0.097	0
PUM50	50	32.9	34.0	31.8	0.952	0.023	0.025	0
PUM155	155	32.9	34.0	31.8	0.886	0.056	0.059	0
PUM215	215	32.9	34.0	31.7	0.797	0.122	0.081	0

soning pointed out above. Considering the increase in intensity on drawing, the origin of the highest temperature peak will be simply attributable to anisotropic fusion process in the time scales of the heating rate of 10 °C/min.

As can be seen from Table 7, although the dried gel does not contain amorphous material, the amorphous phase appears by drawing at 145 °C and could not be removed even if drawn as many as 150 times. The amorphous phase is thought to be removed to produce excellent ultra-high modulus material. If the dried gel is drawn at a slightly lower temperature than 145 °C, the amorphous material may be completely removed. We next examined highly oriented polyethylene samples that were produced by another mode of drawing [78]. A dried gel was obtained by casting the solution of linear polyethylene with a molecular weight of 6×10^6 in an aluminum frame cooled by ice water and drying. It was then solid-state coextruded at 110 °C by the method developed by Kanemoto et al. [79]. This solid-state extrusion corresponds to a 5 fold drawing. It was further drawn until 215 times at the maximum at 135 °C. The phase structure obtained by a similar ^{13}C NMR technique is summarized in Table 8.

As can be seen, the amorphous phase is not present in all samples and the mass fraction of the interphase decreases as the draw ratio increases. By drawing as many as 150 times, the mass fraction of the interphase becomes as low as 0.06. A sufficient effect of drawing is obtained and the elastic modulus becomes as high as 190 GPa.

7
Linear Polyethylene with Randomly Distributed Ethyl Branches (Hydrogenated Polybutadiene)

7.1
Introduction

In the previous sections, we have discussed linear polyethylene samples that were prepared under different conditions. The polymers, called polyethylene, differ widely in molecular structure, i.e. in molecular weight and its distribution, existence of foreign components other than methylene groups and long or short branches. According to such differences in molecular structure, the crystallinity as well as the phase structure in the solid state differ widely. In this section we will examine linear polyethylene that contains a small number of ethyl branches [21].

Examined are two samples that contain 2.2 mol % of ethyl side groups with molecular weights of 16,000 and 400,000 (samples P16 and P420, respectively). The concentration of ethyl branches was determined by standard high-resolution ^{13}C NMR methods [80–82]. These samples were crystallized from the melt by quenching at –78 °C.

7.2
Phase Structure as Revealed from the Analysis of Resonance Lines of Main Chain Methylene Carbons

Figure 18 shows the thermal equilibrium DD/MAS ^{13}C NMR spectra of samples p16 and p420. Two sharp resonance lines are observed at 32.4 and 30.5 ppm, that can be assigned to methylene carbons in the orthorhombic crystalline and noncrystalline phases, respectively. The weak resonance at 10.8 ppm can be assigned to the methylene carbon of the ethyl branches. Examining the longitudinal and transverse relaxation of the methylene resonances at 32.4 and 30.5 ppm, it was found that the crystalline resonance can be represented by a Lorentzian function though it is associated with three different T_{1C} values and the noncrystalline resonance can be broken down into two components with the same T_{1C} and different T_{2C}'s. Based on the relaxation phenomena, a line shape analysis of the equilibrium spectra of both samples was performed. Figure 19 shows the line shape analysis of the equilibrium spectrum of sample P420.

As can be seen, the resonance of the main chain methylene carbons consists of large Lorentzian functions centered at 32.4 and 30.5 ppm and a rather wide Lorentzian function at 32 ppm. These functions represent well the orthorhom-

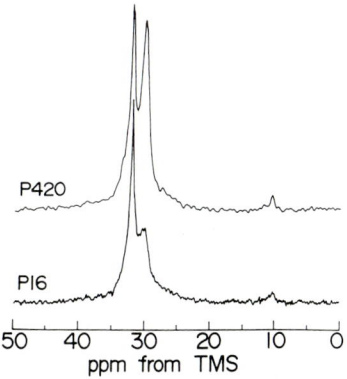

Fig. 18. Thermal equilibrium ^{13}C NMR spectra for two hydrogenated polybutadiene samples P420 and P16, obtained by a τ/2 single pulse sequence (π/4-FID-$τ_\ell$) with $τ_\ell$ = 2000 s

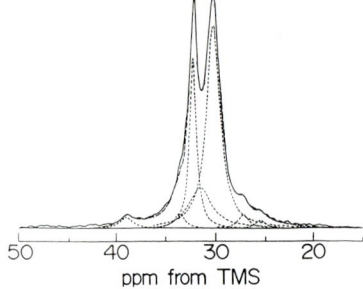

Fig. 19. Line shape analysis of the equilibrium spectrum of P420. The *large dotted Lorentzians* centered at 32.4 and 30.5 ppm and rather *wide dotted Lorentzian* centered at 32 ppm represent the orthorhombic crystalline and noncrystalline amorphous phases and crystalline-amorphous interphase, respectively. The *dotted curve* that is almost completely superimposed on the experimental spectrum indicates the composite curve of the component Lorentzians. *dashed Weakly Lorentzians* at 39, 34, 28, and 26 ppm represent the contributions from the methine and methylene carbons (α and β to the methine and methylene in the ethyl side group), respectively

bic crystalline and noncrystalline-amorphous phases and crystalline-amorphous interphase, respectively. Weak resonances are also observed at 39, 34, 28, and 26 ppm, which can be assigned to the methine and methylene groups (α and β to the methine and methylene in the ethyl side group). These resonances, however, are not pertinent to the present discussion. Thus it is evidenced by the relaxation phenomena and the line shape analysis that the low density linear polyethylene samples examined comprise three phases similar to linear polyethylene; the crystalline, amorphous, and crystalline-amorphous interphase. The mass fraction of each phase was obtained by this line-decomposition analysis. The thicknesses of the three phases were determined from the mass fractions thus obtained with the long spacings determined by the small-angle X-ray diffraction analyses, assuming a stacked lamellar structure. Raman spectroscopic analysis was previously carried out for these samples by Alamo et al. [83]. The phase structure of these samples was also obtained by line shape analysis of Raman spectra based on the elementary spectra of 100% crystalline and amorphous states, and the results are summarized in Table 9.

The half-widths of 37–39 and 78–88 Hz, respectively, for the crystalline and amorphous phases are significantly larger than 18 and 38 Hz for those of the bulk-crystallized linear polyethylene (cf. Table 1). This is caused by incorporation of minor ethyl branches. The molecular alignment in the crystalline phase is slightly disordered, and the molecular mobility in the amorphous phase will therefore be promoted. With broadening of the crystalline and amorphous resonances, the resonance of the interphase also widens in comparison to that of bulk-crystallized linear polyethylene samples. This shows that the molecular conformation is more widely distributed from partially ordered *trans*-rich conformation to complete random conformation, characteristic as the transition phase from the crystalline to amorphous regions.

Table 9. Characterization of phase structure of linear polyethylene with ethyl branches: ^{13}C NMR and Raman spectroscopy

Sample	Parameter	Crystalline	Amorphous	Interphase
P16	chemical shift/ppm	32.4	30.5	32
	Halfwidth/Hz	37	88	133
	mass fraction (NMR)*	0.39	0.43	0.18
	mass fraction (Raman)*	0.37	0.51	0.12
	thickness(A (NMR)*	56	62	13
	thickness/A (Raman)*	53	90	10
P420	chemical shift/ppm	32.4	30.5	32
	halfwidth/Hz	39	78	133
	mass fraction (NMR)*	0.21	0.61	0.18
	mass fraction (Raman)*	0.19	0.68	0.13
	thickness/A (NMR)*	31	88	13
	thickness/A (Raman)*	27	115	10
	T_{1C}/s	82, 11, 0.95	0.16	0.16
	T_{2C}/s milliseconds	0.024	0.41	0.024

* The data from NMR and Raman spectroscopy are taken from Reference [21] and [83], respectively.

It is noted that both the mass fractions and the thicknesses obtained by the NMR method are in very good agreement with the results reported by Alamo et al. using Raman spectroscopy [83]. The thicknesses of the interphases are estimated to be 13Å for two samples, although the molecular weight and the degree of crystallinity differ widely. The results obtained here for the model ethylene copolymers give further support to the concept that there is a quantitatively defined, nonregularly structured interfacial region between the crystalline and amorphous phases.

7.3
Partitioning of Methyl Carbons Between Different Phases

In the equilibrium spectra shown in Fig. 18, the resonance of the methyl carbon in the ethyl branches (referred to as $1B_2$ carbon, hereafter) can be seen in ca. 10 ppm range. It is possible to obtain the information of the partitioning of the ethyl side groups between different phases by analyzing this resonance. In our discussion up to now we have tacitly assumed that the ethyl side groups do not enter the lattice, i.e. the crystalline phase remains pure. A central question associated with this class of copolymers is whether the side group enters the lattice and if so under what conditions, i.e. equilibrium or nonequilibrium. A variety of experimental results, involving many different techniques, have made clear that side groups greater than methyl are effectively excluded from the crystal lattice [84]. Among these many techniques, solid-state ^{13}C NMR methods have been used to study this problem. Although most of the NMR studies agree with the conclusion cited above, there are minor disagreements [85–87]. Studies with a hydrogenated polybutadiene that contained 1.7 mol% ethyl branches led to the conclusion that a very small proportion, 5–10% of the total branch concentrations, was located within the crystalline region. In a study involving polydisperse, broad-composition distribution ethylene-butene copolymers, it was concluded that about 9% of the total branches were located within the lattice [87]. The concentration of ethyl groups entering the lattice that was deduced from these works is admittedly very small. However, since most of these conclusions were deduced without consideration of the presence of the crystalline-amorphous interphase, it is necessary to re-examine the problem by analyzing the resonance of the $1B_2$ carbon.

Figure 20 shows the DD/MAS spectrum in the $1B_2$ resonance range for the sample P16 that was obtained by a single-pulse sequence with a repetition time of 50 s. This spectrum is considered to be the equilibrium $1B_2$ resonance, since the 50 s is longer enough than 5 times the longitudinal relaxation time of the $1B_2$ carbon. There is a recognized downfield broad shoulder in addition to the upfield main resonance centered at 10.8 ppm [88]. It was found by examining the longitudinal and transverse relaxations that both the downfield and upfield resonances are associated with the same T_{1C} of ca. 0.1 s but with different T_{2C}'s; ca. 20 µs for the former, and greater than a few milliseconds for the latter. Based on this analysis, the equilibrium $1B_2$ resonance was broken down into two Lorentzian functions. The result is shown by the dotted curves in the Fig. 20. As can be seen, the composite curve of the two component Lorentzians coincides well with the observed spectrum. The intensity of the downfield resonance is very small relative to the upfield one. The integrated mass ratio of the downfield resonance to the upfield resonance was estimated to be 1:5. Here, the main upfield resonance can be definitely assigned to the $1B_2$ methyl carbons in the noncrystalline amorphous phase. The question remains as to the proper phase assignment of the downfield shoulder. Considering the fact that the T_{2C} is as short as 20 µs, this resonance could be assigned to a branch methyl carbon located either in the crystalline region or in the crystalline-amorphous interphase. The concentration of ethyl branches involved is, however, very small, being in the order of only 0.44 mol%. We note here that the T_{1C} of this component is 0.1–0.2 s, and it is the same as in the amorphous phase.

In principle, a methyl carbon with such a short T_{1C} could be assigned to the crystalline phase if the methyl groups were an integral part of the crystal structure. In

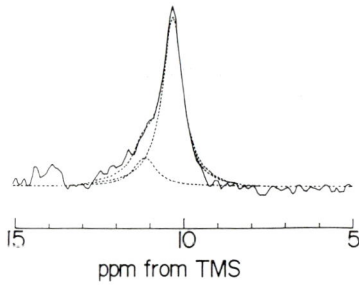

ppm from TMS

Fig. 20. Equilibrium spectrum and line shape analysis in the $1B_2$ carbon range for the sample P16. The spectrum was obtained by a single pulse sequence with a repetition time of 50 s

polypropylene, for example, where this condition is satisfied, the T_{1C} for the methyl carbon in the crystal is much shorter than that for the crystalline methine and methylene ones, as will be described later [24]. The reason for such a short T_{1C} is that there is a threefold rotational motion of the methyl group in a frequency range that has an influence on the spin-lattice relaxation time. However, such an influence would not be expected if the methyl group were not an inherent part of the crystal structure, as is the case of the copolymers being studied here. Hence, based on the very short T_{1C} values, the downfield resonance could be assigned to methyl carbon in the crystalline-amorphous interphase. This assignment is consistent with other studies involving the location of the ethyl side groups [82].

Since a very large proportion of the methyl groups are located in the amorphous region, it is not possible, or reasonable, to assign the downfield resonance completely to the crystalline region. A chain passing between the amorphous and crystalline regions has to traverse the interfacial region. Hence, some of the side groups would have to be located in this region. VanderHart et al. [85, 86] also studied similar copolymers by solid-state ^{13}C NMR. The basic spectra that they reported are very similar to those presented here. In their analysis of a hydrogenated polybutadiene sample containing 1.7 mol% ethyl branches (very similar to the one studied here), they found the ratio of the $1B_2$ carbons of the downfield to upfield resonance to be 1:10. Considering the weak intensity of the downfield resonance, this result, and that of the present work, can be considered to be very similar. However, they assigned the downfield $1B_2$ methyl carbon solely to the crystalline phase because the $T_{1\rho}^H$ of 8.2 ms for the $1B_2$ downfield line almost agreed with the $T_{1\rho}^H$ of 8.7 ms for the crystalline methylene line. Since only two regions, the crystalline and amorphous phases, were considered in their work, it was logical to assign the downfield resonance to the crystalline phase. As was discussed above, with the establishment of a crystalline-amorphous interphase and consideration of the T_{1C} and T_{2C}, the downfield resonance can be assigned to the interphase. The very close $T_{1\rho}^H$ values of the crystalline methylene resonance and the downfield line of the $2B_1$ resonance that were reported can also be understood logically as the downfield line being assigned to the crystalline-amorphous interphase. Thus it is concluded that the ethyl branches are located in the crystalline-amorphous and amorphous phases in a ratio of 1:5. Very minor ethyl branches may be located in the crystalline region; however, it is not possible to estimate such minor concentrations using this NMR technique.

8
Poly(tetramethylene oxide)

8.1
Introduction

Until now we have been concerned mostly with crystalline polyethylene. In this section we consider the solid-state structure of poly(tetramethylene oxide) [22]. Since the melting temperature of this polymer is 42 °C, we examined the structure at temperatures below room temperature. The sample was prepared by ring-opening polymerization of tetrahydrofuran by using triethyloxonium hexafluoroantimonate in methylene chloride at 0 °C. The number-average molecular weight determined by osmometry at 28 °C in toluene, was 67,000. The sample was melted at 65 °C and then crystallized at 20 °C for one month. The heat of fusion and the melting point were estimated to be 8.13 KJ/mol and 42 °C (as the peak temperature) by DSC measurements, respectively. It was confirmed by X-ray diffraction analysis that the crystalline molecular chain of this polymer is in a planar *zig-zag* conformation [89, 90.]

8.2
Experimental

CP/MAS and DD/MAS ^{13}C NMR Spectra. Figure 21 shows the CP/MAS and DD/MAS ^{13}C NMR spectra at 0 °C. This polymer has two different kinds of carbon atom in the monomeric unit, i.e. α- and β-methylene carbons to the oxygen atom. In accordance with this molecular structure, two resonance lines at ca. 72 and 28 ppm are observed in the CP/MAS spectrum (a). These downfield and upfield resonances are assignable to the α and β-methylene carbon atoms, respectively. Here, both resonance lines seem to have upfield shoulders. In order to examine these upfield shoulders, we measured a DD/MAS spectrum by a single pulse sequence without CP. The spectrum (b) shows the spectrum obtained by a single pulse sequence with a repetition time of 5 s. Since 5 s is much shorter than the T_{1C} of the crystalline component, we could see preferentially the noncrystalline contribution that was not clearly recognized in the CP/DD spectrum due to the lesser CP efficiency. In this spectrum, the resonance lines that were recognized in spectrum (a) as only upfield shoulders to the resonance lines of α- and β-methylene carbons are distinctly recognized. Since their chemical shifts are very close to those of the respective carbons of this polymer in CDCl$_3$ solution that are shown in (c) as a stick type spectrum, they are evidently assigned to the noncrystalline component.

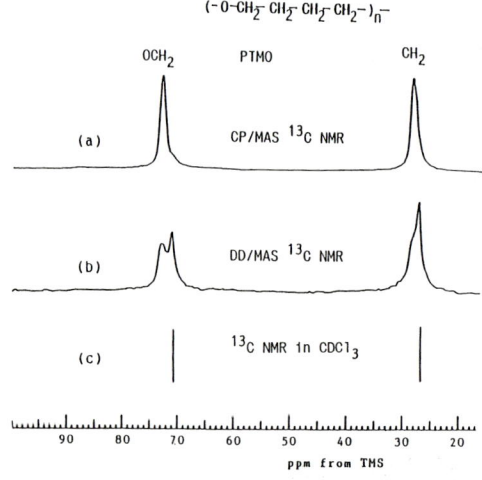

Fig. 21. (*a*) CP/MAS ^{13}C NMR spectrum of poly(tetramethylene oxide) at 0 °C. (*b*) The partially relaxed spectrum. (*c*) Stick-type spectrum in CDCl$_3$ at room temperature

Table 10. Chemical shift, T_{1C} and T_{2C} of α- and β-methylene carbons of poly(tetramethylene oxide)

	Chemical shift (ppm)	T_{1C} (s)	T_{2C} (ms)	Origin
α-CH$_2$	72.9	209, 9.3		crystalline
	70.9	0.15	7.95, 0.099	noncrystalline
	(70.64)[a]			solution
β-CH$_2$	28.3	209, 10.4		crystalline
	27.0	0.14	8.22, 0.099	noncrystalline
	(26.56)[a]			solution

[a] Chemical shift in CDCl$_3$ solution is shown in parentheses.

Spin-Lattice and Spin-Spin Relaxation Times. In an attempt to examine the detail of these noncrystalline resonance lines, T_{1C} and T_{2C} were estimated using a similar method to that used in the previous sections. The results are summarized in Table 10.

The resonance lines at 72.9 and 28.3 ppm are assigned to the crystalline components of α- and β-methylene carbons because of their longer T_{1C} values. These crystalline resonance lines are associated with two T_{1C} values of ca. 209 and 9–10 s. This shows that both methylene carbons possess two components with different T_{1C}'s, but this will not be discussed further, since the existence of plural T_{1C}'s is a normal finding for crystalline polymers as discussed in previous sections. On the other hand, the resonance lines at 70.9 and 27.0 ppm recognized for α- and β-methylene carbons are assignable to the noncrystalline component, because these chemical shifts are very close to those in the solution. These lines are associated with only one T_{1C} of 0.15 or 0.14 s and two T_{2C} values of 7.95 s and 0.099 ms, or 8.22 s and 0.099 ms, respectively for the α- and β-methylene carbons. This suggests that the noncrystalline component involves two components, both associated with a same T_{1C} and different T_{2C}'s. The noncrystalline component with a T_{2C} of 7.95 or 8.22 ms is thought to form an amorphous phase and that with a T_{2C} of 0.099 ms comprises a crystalline-amorphous interphase. In order to confirm this, we examined the elementary line shapes of each component and performed the line shape decomposition analysis of the equilibrium spectrum.

8.3
Phase Structure and Discussion

Figure 22-(a) shows the DD/MAS spectrum in the resonance range of α-methylene carbon at 0 °C. This spectrum represents the thermal equilibrium state of this sample, because it was obtained by a single pulse sequence with the repetition time of 600 s longer than 5 times the longest T_{1C} in the system. The spectrum (b) is that of the crystalline component, which was obtained with use of Torchia's pulse sequence [53]. In the equilibrium spectrum, the noncrystalline contribution (amorphous plus interfacial) can be seen upfield to the crystalline component. Figure 23 shows the elementary line shapes of the amorphous and crystalline-amorphous interphases that comprise the noncrystalline resonance.

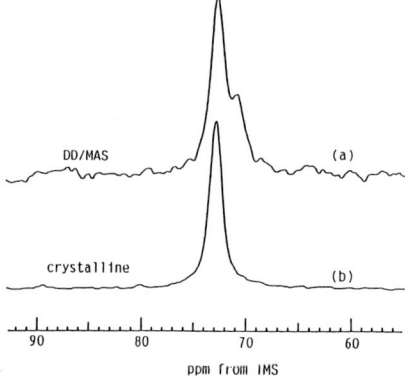

Fig. 22. (*a*) Equilibrium spectrum of poly(tetramethylene oxide) at 0 °C in the range of the α-methylene carbon and (*b*) its crystalline component

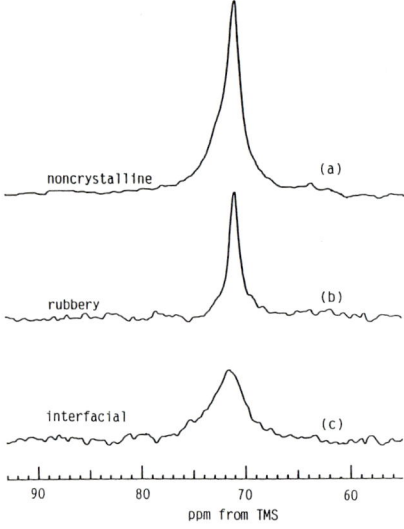

Fig. 23. (*a*) Partially relaxed spectrum of the α-methylene carbon at 0 °C, obtained as a π/2 single pulse sequence with $\tau_\ell = 0.8$ s, and (*b*) transversally relaxed spectrum for 600 μs and (*c*) difference spectrum obtained by subtracting (*b*) from (*a*). Spectrum (*a*) represent all noncrystalline contribution and (*b*) and (*c*) represent the contributions of the amorphous and crystalline-amorphous interphases, respectively

Spectrum (a) shows the DD/MAS ^{13}C NMR spectrum of the α-methylene carbon that was obtained by a single pulse sequence with a repetition time of 0.8 s. This is 0.8 s is longer than 5 times the T_{1C} of the noncrystalline component and much shorter than T_{1C} of the crystalline component (cf. Table 10). Hence, this spectrum represents the contribution from the noncrystalline component that consists of amorphous and crystalline-amorphous interphases. Spectrum (b) is a partially relaxed spectrum transversely for 600 μs. Since 600 μs is much longer than the T^{2C} of 99 μs for the interfacial component, and shorter than that for the amorphous component, this spectrum can be taken as the elementary line shape for the noncrystalline amorphous phase. The elementary line shape for the noncrystalline interfacial phase was constructed by subtracting spectrum (b) from spectrum (a), and is shown as spectrum (c) in the figure. The wide line shape located between the crystalline and amorphous resonances shows that the molecular conformations in this phase are widely distributed over the permitted conformations stationary in time.

The line-decomposition analysis of the equilibrium spectrum of the α-methylene carbon was carried out using the elementary line shapes thus obtained for the three phases. The result is shown in Fig. 24. The composite curve of the decomposed components reproduces well the experimentally observed spectrum. The mass fraction of the crystalline component was estimated as 0.60 that is described in the figure. Based on the heat of fusion of 8.13 KJ/mol of this sample and the value of 14.2 KJ/mol for the crystalline material of this polymer the crystalline fraction was estimated to be 0.57. Here the heat of fusion for the crystalline material was obtained from the effect of diluent on the melting temperature with use of the relationship developed by Flory [91]. The crystalline fraction estimated from the NMR analysis is in good accord with the value estimated from the heat of fusion, supporting the rationality of the NMR analysis.

The mass fractions of the amorphous and crystalline-amorphous interphases are obtained to be 0.18 and 0.22 as described in Fig. 24. For this sample the long spacing was estimated as 145 A° by small-angle X-ray scattering. With use of these mass fractions and the long spacing, the thicknesses of the crystalline, amorphous, and crystalline-amorphous interphases were evaluated to be 87, 26, and 16 A°, respectively, assuming a stacked lamellar crystalline structure. The thickness of the interphase estimated here as 16 A° is in good accord with the value deduced theoretically by Yoon and Flory [7] and the experimental result obtained by small-angle neutron-scattering analysis [38, 92]. Nevertheless, the interfacial

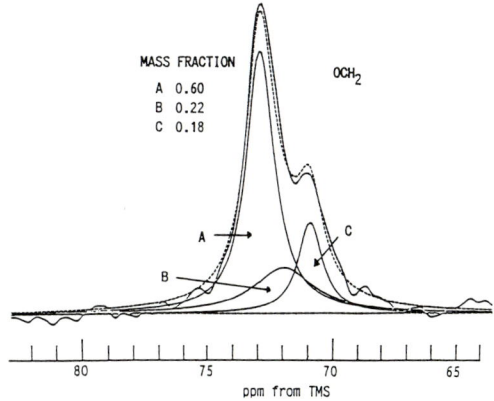

Fig. 24. Line shape analysis of the equilibrium spectrum of the α-methylene carbon shown in Fig. 22-(a). A, B, and C represent the crystalline, crystalline-amorphous interfacial, and amorphous components, respectively

thickness is significantly thinner than that of the bulk-crystallized linear polyethylenes with molecular weights of 20,000–110,000 whose thickness are 34 A° as described in Sect. 3. This implies that the transition region from the crystalline to the amorphous region of this polymer with the molecular chains including the C–O–C bond sequence could be somewhat thinner than that of polyethylene with C–C bonds. The molecular chains with the C–O–C bond sequence may be a little flexible compared to those with only a C–C bond sequence, if both molecular chains comprise planar *zig-zag* chain conformation. This effect may reflect the relatively shorter crystalline T_{1C} value of this polymer in comparison with that of polyethylene. Compare the data in Table 10 with those in Table 1.

9
Isotactic Polypropylene Crystallize from the Melt

9.1
Introduction

Isotactic polypropylene (iPP), which is highly crystalline polymer, has been studied by several groups using solid-state high-resolution ^{13}C NMR spectroscopy. Bunn et al. measured CP/MAS ^{13}C NMR spectra of annealed and quenched α-form (monoclinic) and β-form (hexagonal) samples at room temperature [93]. Doublets with 1:2 intensity were observed of CH_2 and CH_3 resonance lines of the annealed α-form sample. They assumed that these doublets were derived from the presence of two different packing sites in crystals on the basis of the arrangement of the molecular chain in a unit cell of α-form. They also suggested that disappearance of these doublets for quenched α- and β-form samples was due to the decrease in crystalline perfection and the absence of the different sites, respectively.

Lyerla et al. measured T_{1C} over a wide temperature range from room temperature down to 105 K [94], and concluded that T_{1C}'s of not only CH_3 but also CH resonances depend on CH_3 rotational motion, and that the broadening of the CH_3 resonance below –100 °C is also due to modulation of CH_3 rotational motion at the frequency of proton nutation in the presence of the decoupling field. Gomez et al. have also reported solid-state high-resolution ^{13}C NMR spectra of isotactic polypropylenes [95]. They used samples characterized by X-ray crystallography and reconfirmed the results obtained by Bunn et al.

In this section we will discuss the molecular structure of this polymer based on our results mainly from the solid-state ^{13}C NMR, paying particular attention to the phase structure [24]. This polymer has somewhat different character when compared to the crystalline polymers such as polyethylene and poly(tetramethylene) oxide discussed previously. Isotactic polypropylene has a helical molecular chain conformation as the most stable conformation and its amorphous component is in a glassy state at room temperature, while the most stable molecular chain conformation of the polymers examined in the previous sections is planar *zig-zag* form and their amorphous phase is in the rubbery state at room temperature. This difference will reflect on their phase structure.

9.2
Experimental

The iPP sample was used after purification by Soxhlet extraction with toluene. The weight-averaged molecular weight of the purified sample was 2.30×10^5 and the isotacticity was mm = 98%. The bulk-crystallized sample was obtained by isothermal crystallization at 140 °C for 6 days from 230 °C. The crystal structure was confirmed as α-form (monoclinic) by X-ray analysis.

DD/MAS Spectra at Different Temperatures. Figure 25 shows 50 MHz DD/MAS ^{13}C NMR spectra over a wide temperature range together with a stick-type spectrum of the isotactic sequence of this polymer in solution. First of all, for the spectrum at room temperature, three resonance lines assignable to CH_3, CH_2, and CH carbons are visible similarly to those in solution. However, CH_2 and CH resonances of the bulk-crystallized sample exhibit upfield shifts from those in solution, reflecting the difference in conformation between the solid and solution states. Moreover, CH_2 and CH_3 doublets are clearly observed, which are related to the contribution from the crystalline region as reported by Bunn et al. [93]. If these doublets stem from the presence of the different packing sites in a unit cell as they proposed, such a difference might disappear at higher temperatures and the doublets would change to singlet. In fact, with increasing temperature, the doublets of CH_2 and CH_3 resonances tend to singlets, as a result of downfield shift of the upfield component of each doublet. At 147 °C, that is higher than the crystallization temperature of this sample, and a few degrees lower than the melting point, the resonances completely changed to singlets (above 87 °C sharp

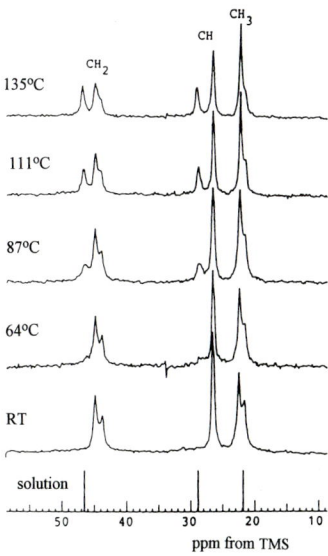

Fig. 25. DD/MAS ^{13}C NMR spectrum of isotactic polypropylene at different temperature with stick-type spectrum in solution

downfield resonance lines appear to the CH_2 and CH carbon resonances, this will be discussed later). However, no change in the spectrum at room temperature was observed even after heating at 147 °C. Moreover, no appreciable change was observed in the X-ray diffraction patterns in the heating and cooling courses in this temperature range. A possible explanation for the change of the NMR spectrum with temperature and no change of the X-ray diffractions could be the packing effect, which causes the CH_2 and CH_3 doublets as proposed by Bunn et al. [93], and by molecular motion in the crystalline regions, which becomes more active at higher temperatures and averages each crystalline carbon resonance over the two sites. This molecular motion is most likely a threefold jump rotation about the 3_1-helical chain axis. Such a motion would produce little change in the X-ray diffraction patterns at higher temperatures as is observed.

At higher temperatures in Fig. 25, sharp resonance lines become visible downfield to the CH_2 and CH resonances as pointed out above. Since their chemical shifts are very close to those of the CH_2 and CH resonances of the isotactic sequence of this polymer in solution, these resonance lines are assigned to the amorphous component. These assignments are also evidenced by examining the spin-lattice and spin-spin relaxation phenomena.

Spin-Lattice and Spin-Spin Relaxation. The T_{1C}'s of CH_2 and CH and CH_3 carbons are summarized in Table 11. For the CH_2 and CH carbons, three different T_{1C}'s are respectively observed, reflecting different molecular mobilities in the crystalline and noncrystalline regions. The longest, shortest, and medium T_{1C} components can be assigned to the crystalline, amorphous, and crystalline-amorphous interphases, respectively, as examined below in detail. However, only two T_{1C} values are recognized for the CH_3 carbon and the chemical shift is very close to that of the CH_3 carbon in the isotactic sequence in solution. The reason for this is that there is a threefold rotational motion of the CH_3 group throughout the crystalline and amorphous regions as in solution. The longer and shorter T_{1C}'s (1.31 and 0.47 s) can undoubtedly be assigned to the crystalline and amor-

Table 11. Chemical shift and ^{13}C spin-lattice relaxation times of bulk isotactic polypropylene at 87 °C

	Chemical shift/ppm	T_{1C}/s	Origin
CH_2	46.43	0.62	amorphous
	44.82	69.1	crystal
		7.1	interface
CH	28.76	0.32	amorphous
	26.60	47.1	crystal
		4.9	interface
CH_3	22.39	1.31*	crystal
		0.47*	amorphous

* T_{1C} of CH_3 carbon for the crystalline-amorphous interphase could not be decided, either 1.31 or 0.47 s.

phous components, respectively. The T_{1C} of 1.31 s is significantly shorter than T_{1C} for the other carbons such as CH and CH_2 assigned to the crystalline component. This is because there is a threefold rotational motion of the CH_3 group even in the crystalline region as pointed out by Lyerla et al. [94]. In addition to the crystalline and amorphous phases, the presence of the crystalline-amorphous interphase was ascertained as described later. What is the T_{1C} for the methyl carbon in this phase? It should be 1.31 or 0.47 s; that is, the former if the threefold rotational motion occurs as in the crystalline phase and the latter if it occurs as in the amorphous phase. It may be important to decide the degree of the threefold motion of methyl group in the crystalline-amorphous interphase. However, we could not reach any conclusions only using the T_{1C} results.

9.3
Phase Structure and Discussion

In an attempt to investigate the phase structure of this sample, the line shape analysis of the CH_2 resonance line in the DD/MAS spectrum at 87 °C that is shown in Fig. 25 was examined. The result is shown in Fig. 26-(a). The elementary line shape of the crystalline phase was obtained as the line shape of the longest T_{1C} component by Torchia's pulse sequence [53]. It was a doublet and was represented approximately by two down- and upfield Lorentzians with an intensity ration of 2:1 (Spectrum A shown by dotted line in Fig. 26). Since all methylene carbons in the α-crystalline form of this polymer are equivalent in the intramolecular helical conformation, the origin of the doublet could be attrib-

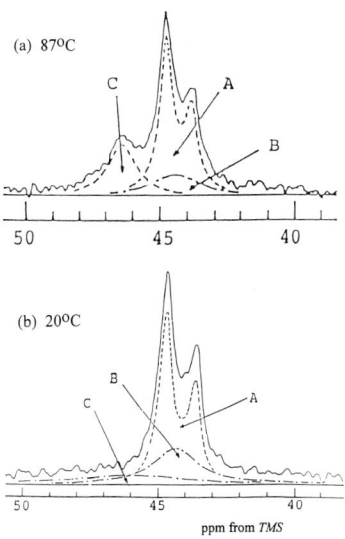

Fig. 26. Line shape analysis of the CH_2 resonance line of bulk iPP at 87 and 20 °C. A, B, and C indicate the crystalline, crystalline-amorphous interfacial, and amorphous components, respectively

uted to an intermolecular packing effect. In fact, the doublet with the ration of 2:1 can be well understood as having originated from the existence of two packing sites with different magnetic environments, examining the unit cell of the α-crystalline form. The elementary line shape of the amorphous phase was obtained using the spin-lattice and spin-spin relaxations as a component with T_{1C} of 0.62 s (cf. Table 11). This elementary line shape, shown by a dotted line as Spectrum C in Fig. 26, could be well approximated by a Lorentzian centered at 46.43 ppm. The relatively narrow line width and the chemical shift close to that of the CH_2 resonance line of this polymer in solution show the rubbery amorphous state of this component.

The line shape analysis of the resonance of the CH_2 carbon at 87 °C was carried out with use of the elementary line shapes thus obtained for the crystalline and amorphous components. However, well reproducible results as shown by the dotted lines in Fig. 26-(a) were obtained only when an additional broad Lorentzian (spectrum B in the figure) was introduced. The broad component B introduced here corresponds to the component with T_{1C} of 7.1 s at a chemical shift of 44.82 ppm shown in Table 11. Considering its broad line width and T_{1C}, this component can be assigned to the crystalline-amorphous interphase. However, the chemical shift is very close to that of the crystalline phase and the T_{1C} is as long as 7.1 s. For most crystalline polymers examined in the previous sections, the T_{1C} of the crystalline-amorphous interphase is normally as short as a few tenths of a second, that is the same as the amorphous phase. On the contrary, the T_{1C} of 7.1 s assigned here to the interphase is appreciably longer than the 0.62 s of the amorphous phase. Such a character of the interphase is also recognized for the CH resonance as can be seen in Table 11. For the CH carbon, the T_{1C} of 4.9 s assignable to the interphase is appreciably longer than the 0.32 s assignable to the amorphous phase. Taking into consideration the coincidence of the chemical shift with that of the crystalline phase, the reason for this will be that the helical molecular chain conformation is still held to some extent even in the crystalline-amorphous interphase and the molecular motion in the T_{1C} frequency range is somewhat restricted in comparison with the amorphous phase.

Similar line shape analyses for the equilibrium spectra at different temperatures were performed. At room temperature, where the amorphous phase is in a glassy state, the determination of the elementary line shape of the amorphous component was a little difficult. However, excellent line-decomposition analysis was obtained by introducing a broader Lorentzian centered at the same chemical shift as at higher temperatures. The result at room temperature is shown in Fig. 26-(b). Here the nature of the component line shapes A and B of the crystalline and crystalline-amorphous interphases is similar to that in the spectrum at 87 °C. However, the component line shape for the amorphous component is quite different from that at 87 °C that is distributed over a very wide chemical shift range centered at the same chemical shift to that at higher temperatures. This reflects the glassy state of the amorphous phase. In the glassy state, the molecular conformation in the amorphous phase will be distributed over all permitted conformations stationary in time and randomly in space. The wide component line shape of the amorphous component obtained here at room temperature well represents this molecular nature of the amorphous phase.

Table 12. Mass fractions of three phases and T_{1C} obtained by analysis of CH_2 resonance line at different temperatures

Temperature (°C)	Mass fraction			T_{1C}/s		
	Crystalline	Amorphous	Interphase	Crystalline	Amorphous	Interphase
23	0.57	0.16	0.27	65	–*	7.5
64	0.57	0.18	0.26	69	0.18	7.5
87	0.57	0.26	0.18	70	0.20	7.0
110	0.57	0.33	0.10	60	0.21	7.0

* T_{1C} of the amorphous phase could not be decided because of the broad line width of the resonance lines for amorphous and interfacial phases at 23 °C. However, it is supposed to be ca. 0.2 s.

The mass fraction of each phase that was obtained by the line shape analysis of the CH_2 resonance line at different temperatures is summarized in Table 12 with T_{1C}. It can be seen that the mass fraction of the crystalline phase (degree of crystallinity) stays unchanged at 0.57 over the wide temperature range from room temperature to 110 °C, while the amorphous phase increases and the interphase decreases with increasing temperature. The T_{1C} of the CH_2 carbon in each phase is mostly unchanged over the temperature range examined; 65~70 s for the crystalline phase, 0.18~0.21 s for amorphous phase, and 7.0~7.5 s for the interphase. This shows that the molecular motion of each phase in the T_{1C} time frame is almost the same either in the glassy or rubbery state.

As pointed out above with relation to the data at 87 °C, the T_{1C} of the crystalline-amorphous interphase is appreciably longer than that of the amorphous phase, suggesting the retention of the helical molecular chain conformation in the interphase. We also note that a T_{1C} of 65~70 s for the crystalline phase is significantly shorter than that for other crystalline polymers such as polyethylene and poly-(tetramethylene oxide), whose crystalline structure is comprised of planar *zig-zag* molecular-chain sequences. In the crystalline region composed of helical molecular chains, there may be a minor molecular motion in the T_{1C} frame, with no influence on the crystalline molecular alignment that is detected by X-ray diffraction analyses. Such a relatively short T_{1C} of the crystalline phase may be a character of the crystalline structure that is formed by helical molecular chain sequences.

10
Syndiotactic Polypropylene Gel

10.1
Introduction

We studied the structure of bulk-crystallized isotactic polypropylene in the previous section, in this section we will examine the gel structure of syndiotactic polypropylene (sPP) [25]. Many crystalline polymers form a gel from the concentrated solution. The network entities that hold the gel structure are reported as crystallites for crystalline polymers such as polyvinyl alcohol [96, 97] and polystyrene [98, 116]. The entities that form sPP gel must also be crystallites. The crystal structure and molecular conformation of this polymer in the bulk state have been widely studied by X-ray and electron diffraction analyses [99–109], solid-state high-resolution ^{13}C NMR [110–113] and vibrational spectroscopy [114–116]. With relation to the molecular conformation, three types of helical structure have been reported. The most stable form consists of a *trans-trans-gauche-gauche* molecular sequence (referred to as a *ttgg* form), but a metastable form with all *trans* molecular sequences (*tttt* form) is also produced by cold drawing. In addition to these two forms, the formation of a *ttggttttttgg* sequence has recently been reported when the cold drawn film is exposed to a solvent vapor such as toluene [108]. To elucidate the molecular conformations of the bulk-crystalline sPP, solid-state high-resolution ^{13}C NMR spectroscopy has been found very effective, since the spectrum reflects sensitively the molecular conformation. In this section, we will examine the molecular conformation of sPP gel, referring to the above-cited molecular conformations in the bulk state.

10.2
Experimental and Discussion

A sPP sample with a z-averaged molecular weight of 82.2×10^6 was used. Its racemic triad was estimated to be 95%. The gel sample was obtained by quenching a 13.6 wt% o-dichlorobenzene solution of the sPP in ice water at 150 °C.

Molecular Conformation of sPP gel. Figure 27 shows the DD/MAS ^{13}C NMR spectrum of sPP gel. This spectrum was obtained by a single-pulse sequence $(\pi/2–\text{FID}_{DD}–\tau_\ell)_n$ with the repetition time τ_ℓ more than 5 times the longitudinal relaxation time T_{1C}. Hence, this spectrum reflects the thermal equilibrium state of the gel. For comparison, the spectrum of the bulk *ttgg* crystal of this sample

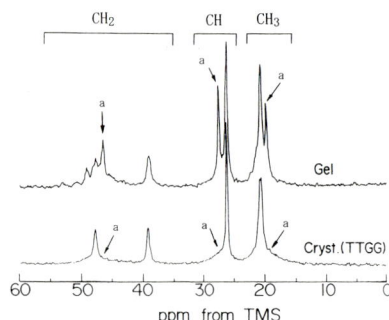

Fig. 27. Equilibrium DD/MAS ^{13}C NMR spectrum of sPP/*o*-dichlorobenzene gel (13.6 wt%), obtained by a π/2 single pulse sequence with a repetition time of 300 s. In the *lower part* the equilibrium spectrum of the bulk sPP crystal in *ttgg* form is shown for reference. The arrows indicate the resonance of the amorphous component of each carbon

is also shown in the lower part of the figure. At first glance the spectrum for the gel is significantly different to that of the bulk crystal. Several additional resonance lines are recognized in the gel spectrum that are not recognized in the *ttgg* crystal. We have to assign these resonance lines to obtain information about the molecular conformation and phase structure of sPP in the gel form. All resonance lines in this spectrum are assigned to polymer carbons, because the resonance lines from the carbons of the solvent, o-dichlorobenzene, are thought to appear in the range of ca. 130 ppm. Referring to the spectrum of this polymer in solution, the resonance lines at ca. 20 ppm are assigned to the methyl carbon, and those in the ranges of 25–30 ppm and 35–55 ppm to the methine and methylene carbons, respectively.

To provide further detailed assignment of these lines, we examined the longitudinal and transverse relaxation times. The results are summarized in Table 13. The CH_3, CH, and CH_2 resonance lines visible for the gel sample at 19.9, 27.4, and 46.4 ppm, respectively, are associated with T_{1C}'s of 0.3, 0.2, and 0.2 s and T_{2C}'s of 14, 11, and 13 ms, respectively. The T_{1C} and T_{2C}'s recognized here are typical of those for the rubbery state of polymers and their chemical shifts are very close to those recognized in sPP solution. Hence, these resonance lines are assigned to the amorphous component of sPP in the gel form. We note that these lines are visible in the gel as sharp lines whereas they are only detectable in the spectrum of the bulk crystal as shoulders to the crystalline lines. The reason for this is that the gel contains a rubbery amorphous component whereas the amorphous component in the bulk crystal is in the glassy state. Since in the rubbery state the molecular conformations are rapidly changed over all permitted conformations, the resonance line becomes narrow and visible. On the contrary, since the molecular conformations in the glassy state are mostly fixed over permitted conformations stationary in time and randomly in space, the resonance line widens and becomes invisible.

Table 13. Chemical shift and T_{1C} values of sPP/o-dichlorobenzene gel (13.6 wt%) and bulk *ttgg* crystal

	ttgg Crystal			Gel			
	Shift (ppm)	$T_{1C}(s)$	$T_{2C}(ms)$	Shift (ppm)	$T_{1C}(ms)$	$T_{2C}(ms)$	Origin
CH_3				19.9	0.3	14	amorphous
	20.7*	0.2	0.077	20.6	0.3	0.052	crystal
	20.9*	0.2	0.077				crystal
CH	26.2	30	0.022	26.2	23	0.010	crystal
				27.4	0.2	11	amorphous
CH_2	39.1	63	0.014	39.0	58	0.012	crystal
				46.4	0.2	13	amorphous
	47.7	63	0.014	47.7	53	0.015	crystal
				49.0	53	0.016	crystal

* The splitting of the methyl resonance is reported in detail in ref. [111].

On the other hand, other lines are assignable to crystalline carbons, because these lines are generally associated with a relatively long T_{1C} and very short T_{2C} that are identical to those of the *ttgg* crystal. The CH_3 resonance line at 20.6 ppm assigned to the crystalline carbon is associated with an exceptionally short T_{1C} of 0.3 s. This is caused by the threefold rotational motion of the methyl group that is permitted even in the crystlline region as discussed with relation to the T_{1C} of methyl carbon of the bulk-crystallized iPP in the previous section. Here some significant difference is noted in the crystalline CH_2 resonance lines between the *ttgg* crystal and the gel. For the *ttgg* crystal two split lines are recognized for the methylene carbon at 39.0 and 47.7 ppm. This split can be the result of the γ-*gauche* effect. It is semiempirically confirmed that the adjacent methine carbon separated by three bonds at the *gauche* position shifts the methylene resonance by ca. 5 ppm upfield [109–111]. Figure 28 shows the helical molecular structure of the *ttgg* crystal of sPP. In the figure, the side view and the view from the c axis are shown. The closed and open circles in the main chain indicate the methylene and methine carbons, respectively. In the *ttgg* molecular sequence

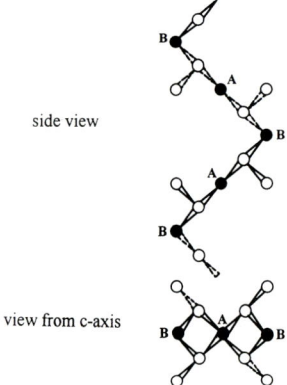

Fig. 28. Molecular structure of the *ttgg* crystal form of sPP. *a* and *b* show the relationship between the methylene and methine carbons separated by three bonds, in *trans* (**a**) and *gauche* (**b**). *c* shows the side view and the view from the c axis of the molecular sequence in the *ttgg* crystal form. The *closed* and *open circles* in the main chain show the methylene and methine carbons, respectively. The methylene carbons marked as *A* and *B* are the methylene carbon with two *gauche* effects (the *gt-gt* carbon) and that without *gauche* effect (the *tg-gt* carbon), respectively. See the text

there are two kinds of methylene carbons as can be seen in the figure; one has tow *gauche* effects and the other no *gauche* effect. We refer to the former as the *gt-tg* methylene carbon and the latter as the *tg-gt* methylene carbon hereafter. In the figure, the methylene carbon indicated as A is the gt-tg methylene carbon and that indicated as B is the *tg-gt* methylene carbon.

The peak at 47.7 ppm is assigned to the methylene carbon of the *tg-gt* sequence without such *gauche* methine carbons in the *ttgg* crystal, whereas the upfield 39.0 ppm line is assigned the *gt-tg* methylene carbon with two *gauche* methine carbons in the ttgg form. On the other hand, three distinguishable lines are recognized at 39.0, 47.7 ppm, and 49.0 ppm to the crystalline methylene carbon in the gel. The intensity ration of the three lines is 3:2:1. The chemical shifts of the former two, 39.0 and 47.7 ppm, are the same as those of the *ttgg* crystal. They are assignable to the *gt-tg* and *tg-gt* methylene carbons in the *ttgg* form as in the bulk crystal. The chemical shift of 49.0 ppm is identical to that of the methylene carbon in the metastable *tttt* form that is observed in cold-drawn sPP. However, this line could not be assigned to the *tttt* crystal as discussed below. As pointed out at the beginning of this section, there is one more molecular chain sequence form, i.e., the *ttggttttttgg* form for this sPP polymer. The methylene chemical shift of this sequence could split into three, corresponding to the methylene carbons in the *tt-tt, gt-tg,* and *gt-tt* sequences. However, for the *ttggttttttgg* crystal, a resonance line for the methylene carbon should appear at 44 ppm from the *gt-tt* sequence with one γ-*gauche* effect. Since no resonance is recognized at 44 ppm, the triple splitting of the methylene resonance of the gel cannot be attributed to the existence of the *ttggttttttgg* crystal.

Since two resonance lines at 39.0 and 47.7 ppm that correspond to those observed in the *ttgg* form and a resonance line at 49.0 ppm that corresponds to that in the *tttt* form are recognized in the gel spectrum, a coexistence of these two forms in the gel might be supposed. In an attempt to determine the possibility of the coexistence of the two forms in the gel, we measured the IR spectrum that is sensitive to the molecular conformation. The number of normal vibrational modes depends sensitively on the molecular conformation based on the selection rule of the symmetry species. Kobayashi et al. confirmed the vibrational modes assignable to the *ttgg* conformation in the IR spectrum for the gel from a sPP/carbon disulfide system [117]. However, since we used o-dichlorobenzene as solvent, we examined whether the gel structure depends on the solvent.

Figure 29 shows the IR spectra of the *tttt* and *ttgg* crystals and the gel. Here the spectrum of the gel was obtained by subtracting the contribution from o-dichlorobenzene. The differences in the number of detected IR active bands between the *tttt* and *ttgg* crystals reflect the number of monomeric units contained in the fiber period as well as the factor group symmetry. It can be seen that all vibrational modes of the gel are completely the same as the ttgg crystal. This indicates that the molecular conformation in the gel takes a ttgg sequence as reported for sPP/carbon disulfide gel [116]. The absorption bands characteristic of the tttt crystal, such as those at 1131 and 830 cm-1, cannot be recognized in the gel spectrum. This distinctly evidences that the crystal molecular conformation in the gel takes the *ttgg* form without coexistence of the *tttt* forms. This conclusion

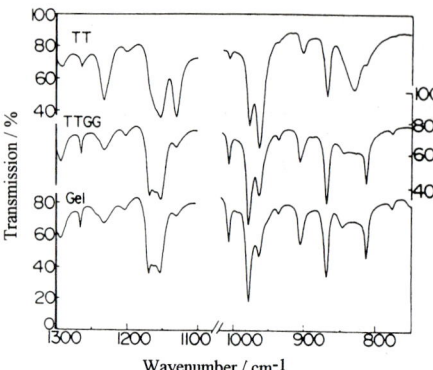

Fig. 29. IR spectra of the *tttt* and *ttgg* bulk crystals and sPP/*o*-dichlorobenzene gel (13.6 wt%). The spectrum of the gel represents only the contribution from the sPP polymer that was obtained by subtracting the contribution from the solvent

is also supported by recent ^{13}C NMR experiments [118]. The triple splitting of the resonance line of the methylene carbon in the ^{13}C NMR spectrum was examined at high temperatures. It was confirmed that the resonance line at 49.0 ppm stays essentially unchanged upon elevating the temperature until melting. Since the *tttt* form is unstable at temperatures above 50 °C, this result also supports the conclusion mentioned above. Therefore, all splitting lines of the methylene carbon in the ^{13}C NMR spectrum can be assigned to the *ttgg* sequence.

Referring to the intensity ratio of 3:2:1 for resonance lines at 39.0, 47.7, and 49.0 ppm, it is concluded that the resonance of the methylene carbon in the crystalline *ttgg* sequence in gel splits into two lines assignable to the methylene carbons with two γ-gauche effects and without γ-gauche effect in the ratio of 1:1, the former at 39.0 ppm and the latter further splits into 47.7 and 49.0 ppm. The intensity ratio of 3:2:1 for lines at 39.0, 47.7 and 49.0 ppm is thus plausibly understood. The splitting of the resonance line assignable to the methylene carbon without a γ-gauche effect into 47.4 and 49.0 ppm cannot be explained by a difference in the molecular sequence conformation. A possible explanation for splitting is given by taking a molecular packing effect into account. The resonance assigned to the *tg-gt* methylene carbon further splits into 47.7 and 49.0 ppm due to the molecular packing effect while the resonance from the *gt-tg* carbon stays as a singlet at 39.0 ppm. Sozzani et al. observed a peak at 49.0 (50.2 ppm estimated by them) in an as-polymerized sPP sample [111]. They did not assign it to the tt-tt conformation. Since the *tg-gt* methylene is located at the outer position of the helical structure of sPP, an enhanced interaction with neighboring helical molecular chains is expected. Therefore, there will be two distinct magnetic environments possible for the *tg-gt* methylene to yield the splitting. The splitting width of 1.3 ppm observed here for the gel is comparable to that in the α-crystal of iPP where a splitting of 1.0 ppm has been reported as described in the previous section [24]. This splitting is understood as being due to the existence of inequivalent sites for methylene carbons due to the molecular alignment in the crystal lattice. In the case of the sPP gel, the same effect is predicted.

Phase structure. It was confirmed in the previous section that the bulk iPP crystal consists of three phases; the crystalline, noncrystalline amorphous phase and crystalline-amorphous interphase. Hence, it is also assumed that the bulk sPP crystal forms a three-phase structure in a similar manner. The question here is whether the sPP crystal involves such a phase structure in forming a gel or not? In order to study this problem, we have analyzed ^{13}C NMR spectra of the sPP gel. The noncrystalline contributions to each resonance of CH_2, CH and CH_3 carbons in the DD/MAS ^{13}C NMR spectrum of the gel can be seen, as indicated by the arrows in Fig. 27, where their assignment as noncrystalline resonances was confirmed by the spin-lattice and spin-spin relaxation times as described above with relation to the results in Table 13. We carried out the line-decomposition analysis of the resonance lines of the methine and methyl carbons, since these resonances are most pertinent for the present purpose because of the simplicity of the spectral shape.

Figure 30 shows the component analysis of the resonances of the methine and methyl carbons in the equilibrium DD/MAS ^{13}C NMR spectrum. Here a Lorentzian function is assumed for each component. The rationality for this assumption was confirmed by examining the elementary line shapes for each component using the differences in the T_{1C} and T_{2C} values in a similar way to that described in preceding sections. The narrow Lorentzian components centered at 26.2 and 20.6 ppm, and 27.4 and 19.9 ppm are assignable to the methine and methyl carbons in the crystalline and amorphous phases, respectively, as discussed previously (see Table 13). In addition to these components, broad Lorentzian components are recognized centered at 26.6 and 21.1 ppm for the methine and methyl carbons. It was

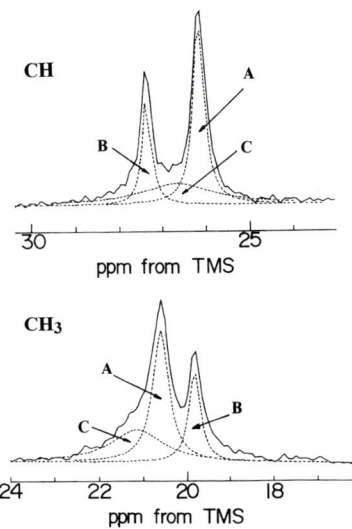

Fig. 30. Line shape analyses of the resonance lines of methine and methyl carbons in sPP/o-dichlorobenzene gel. *A*, *B*, and *C* indicate the crystalline, amorphous, and crystalline-amorphous interfacial components, respectively. (This figure was obtained by revising Fig. 7 in Ref. 25 whose horizontal chemical shift axis was somewhat shifted)

Table 14. Mass fractions of three phases of sPP polymer in sPP/o-dichlorobenzene gel (13.6%)

	Crystalline phase	Amorphous phase	Interphase
CH	0.43	0.19	0.38
CH$_3$	0.44	0.22	0.34

confirmed by minute examination of their longitudinal and transverse relaxations that these components are associated with T$_{1C}$ values as short as 0.2–0.3 s and very short T$_{2C}$ values comparable to those of the crystalline phase. In addition, these broad Lorentzians are widely distributed in a close range of the crystalline component of each carbon. Hence, these broad components can be assigned to the methine and methyl carbons in the transition phase from the crystalline to amorphous. In this phase the molecular conformations are widely distributed over the permitted conformations. The narrow line width of the amorphous components of both the CH and CH$_3$ carbons indicates that the molecular conformations are rapidly interchanged between the permitted conformations in the amorphous phase in the sPP crystal to form the gel. We conclude that the sPP crystal of the gel involves three phases, the crystalline, amorphous, and an interphase.

The mass fractions of these three phases are shown in Table 14. the crystalline fraction is relatively small as 0.43 or 0.44. This low level of crystallinity may arise from relatively strong molecular entanglement due to the network structure.

Phase Structure as Revealed from the Mobility of the Solvent. The phase structure of the sPP crystal in the gel form, which was elucidated by the line-decomposition analysis of the DD/MAS ^{13}C NMR spectrum, will reflect on the mobility of the solvent in the gel. The mobility of the solvent can be examined by the longitudinal relaxation of resonance lines assigned to the carbons of the solvent. Figure 31 shows the longitudinal relaxation for the line at 130 ppm of the o-dichlorobenzene. The open circles indicate the data of the pure solvent and the closed ones those of the solvent in the gel. As can be seen, the relaxation of the pure solvent evolves exponentially with a T$_{1C}$ of 3.0 s, whereas that of the solvent in the gel evolves nonexponentially. This indicates that there are some solvent molecules in the gel that differ in their mobility. We assume here that the longitudinal relaxation of each component of the solvent evolves exponentially. Then the longitudinal relaxation of the total solvent follows the relationship:

$$I = \sum_i I_{0i} \left[1 - \exp(-\tau/T_{1i})\right]$$

Here τ is the evolution time, I denotes the total longitudinal magnetization of the solvent at τ and I_{0i} the longitudinal magnetization of i-th component at $\tau = \infty$. T_{1i} is the spin-lattice relaxation time that dictates the evolution of the magnetization of the i-th component. The Σ_i I_{0i} is the eqilibrium magnetization of the total solvent at $\tau = \infty$. Since the equilibrium magnetization of each component is proportional to the mass fraction, I_{0i} /Σ_i I_{0i} provides the mass fraction of the i-th component.

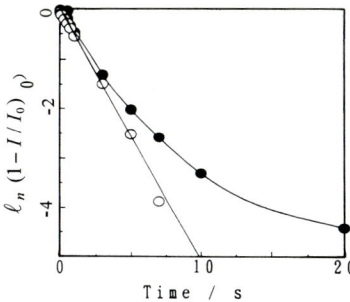

Fig. 31. ^{13}C longitudinal relaxation of the 130 ppm resonance line of o-dichlorobenzene. ● indicates data in gel and ○ data in the pure solvent

Table 15. Mass fractions of the bound and free solvent and those of the crystalline-amorphous interphase and amorphous phase in the noncrystalline tow phases of the polymer in the gel

	Mass fraction	
	Bound solvent (interphase)	Free solvent or (or amorphous)
solvent[a]	0.70	0.30
polymer[b]	0.64	0.36

[a] The mass fractions of bound and free solvents determined from Eq. (4).
[b] The average of the values from CH and CH_3 data in Table 13.

Using this relationship the longitudinal relaxation of the solvent shown in Figure 31 was analyzed. The longitudinal relaxation could be well fitted as the sum of the relaxations of two components, the relaxation of each is dictated with the T_{1C}'s of 2.2 and 9.2 s. It is known that in the gel of a polymer/solvent system the solvent can be generally classified into two components that differ in mobility, referred to as free and bound solvents. The two T_{1C}'s of 2.2 and 9.2 s obtained here correspond to those for these free and bound solvents. The shorter T_{1C} of 2.2 s is almost equal to that of pure o-dichlororbenzene. Hence, the other component with T_{1C} of 9.2 s can be assigned to the bound solvent. The mass fractions of these two components of the solvent that are obtained from $I_{0i}/\Sigma_i I_{0i}$ are listed in Table 15, together with the mass fractions of the amorphous and interphase of the sPP polymer in the gel.

As can be seen from the Table 15, 70 and 30% of the solvent are respectively in the bound and free state. The mass fractions of the bound and free solvents are in rough accordance with those of the crystalline-amorphous interphase and the amorphous phase in the two noncrystalline phases of the polymer. This result suggests that the solvent exists in the two noncrystalline phases of the polymer, as the bound solvent in the crystalline-amorphous interphase and as the free solvent in the amorphous phase, leaving the crystalline phase pure. It is concluded that the sPP/o-dichlorobenzene gel involves three phases, (1) the pure crystalline phase of the polymer, (2) the amorphous phase which consists of amorphous polymer molecules and free solvent, (3) the crystalline-amorphous interphase which consists of polymer molecules and bound solvent.

In the amorphous phase, the molecular conformation and mobiltiy of polymer are similar to those in the rubbery state of the pure polymer and the mobility of the solvent is almost the same as in the pure state. In the crystalline-amorphous interphase, the molecular conformation and mobility of the polymer are severely restricted even if it contains the solvent molecules. Since the solvent molecules are trapped among polymer molecular chains in the interphase, the fluidity as a liquid is restricted. The molecular chain entanglement of the polymer in the interphase plays an important role in preventing the solvent molecules from flowing out of the network structure. A high percentage of the interphase region will be essential to hold the solvent in the network. On the other hand, the solvent molecules in the amorphous phase of the polymer are mostly free from surrounding polymer molecules. They could have a fluidity as pure solvent and could be segregated from the gel with ease. If the polymer chain does not exist to a significant extent in the interphase region, gelation will not occur since most of the solvent molecules are in a free state. Alternatively, the aggregation of polymer molecular chains will lead to a precipitation such as crystal growth from dilute solution.

11
Concluding Remarks

We have reviewed briefly our studies of the solid-state structure of some crystalline polymers. It has been found that bulk-crystallized linear polyethylene involves three phases that are distinctly different from each other in molecular conformation and mobility; the crystalline, the noncrystalline amorphous phase and the crystalline-amorphous interfacial phase. Such a three-phase structure is essentially universal for many semicrystalline polymers. In the case of polyethylene, the amorphous phase can be absent for the solution-grown sample or ultra high modulus sample (high modulus polyethylene). However, even if crystallized from the melt at high pressure and high temperature or highly drawn at high temperatures under normal conditions, the samples generally involve a similar three-phase structure. In addition to polyethylene, we have briefly reviewed the study of poly(tetramethylene oxide) and polypropylene that, respectively, have a planner *zig-zig* and helical molecular chain conformation. It was found that these polymers involve essentially a similar three-phase structure, characteristic of the molecular structure, as well as the thermal and mechanical history of samples. Even the sPP crystal that forms a gel network exhibits a phase structure involving a crystalline-amorphous interphase.

The study of the phase structure of semicrystalline polymers, particularly that of the interfacial region, is of foremost importance, since it has an intimate relationship with the macroscopic properties of polymers. Many techniques have been used to study this problem. To determine the mass fraction of each phase of the three-phase structure, such techniques as ^{1}H broad-line NMR or Raman spectroscopy may be useful. However, these methods are not applicable generally to crystalline polymers other than polyethylene and they cannot be used to characterize the phase structure in a molecular level. For example, in the Raman spectroscopy the elementary line shape of each phase could not be obtained with ease. For polyethylene, the elementary line shapes of the crystalline and amorphous components could be determined referring to the line shapes of the 100% crystalline and molten states [32]. However, it is not possible to obtain such elementary line shapes for other polymers. As the result there are some conflicting conclusions, particularly about the crystalline-amorphous interphase [119].

On the other hand, in the solid-state high resolution ^{13}C NMR, elementary line shape of each phase could be plausibly determined using magnetic relaxation phenomenon generally for crystalline polymers. When the amorphous phase is in a glassy state, such as isotactic or syndiotactic polypropylene at room temperature, the determination of the elementary line shapes of the amorphous and crystalline-amorphous interphases was not so easy because of the very broad line width of both the elementary line shapes. However, the line-decomposition analysis could plausibly be carried out referring to that at higher temperatures where the amorphous phase is in the rubbery state. Thus, the component analysis of the spectrum could be performed and the information about each phase structure such as the mass fraction, molecular conformation and mobility could be obtained for various polymers, whose character differs widely.

References

1. Flory PJ (1949) J Chem Phys 17: 223
2. Flory PJ (1962) J Am Chem Soc 84: 2857
3. Mansfield ML (1983) Macromolecules 16: 914
4. Marqusee JA, Dill KA (1986) Macromolecules 19: 2420
5. Marqusee JA (1989) Macromolecules 22: 472
6. Flory PJ, Yoon DY, Dill KA (1984) Macromolecules 17: 862
7. Yoon DY, Flory PJ (1984) Macromolecules 17: 868
8. Kumar SK, Yoon DY (1989) Macromolecules 22: 3458
9. Suniga I, Rodrigues K, Mattice WL (1990) Macromolecules 23: 4108
10. Dill KA, Naghizadeh J, Marqusee JA (1988) Ann Rev Phys Chem 39: 425
11. Leemaker FAN, Scheutjens JMHM, Gaylord R (1984) Polymer 25: 1577
12. Mandelkern L (1992) Chemtracts-Macromol Chem 3: 347
13. Bergmann K, Nawotky K (1967) Koloid-Z Z Polym 219: 132
14. Bergmann K (1973) Koloid-Z Z Polym 251: 962
15. Kitamaru R, Horii F, Hyon S-H (1977) J Polym Sci Polym Phys Ed 15: 821
16. Kitamaru R, Horii F (1978) Adv Polym Sci 26: 137
17. Kitamaru R, Horii F, Murayama K (1986) Macromolecules 19: 636
18. Kitamaru R, Horii F, Zhu Q, Bassett DC, Olley RH (1994) Polymer 35: 1171
19. Nakagawa M, Horii F, Kitamaru R (1990) Polymer 31: 323
20. Zhu Q, Horii F, Tsuji M, Kitamaru R (1989) J Soc Rheology, Japan, 17: 35
21. Kitamaru R, Nakaoki T, Alamo RG, Mandelkern L (1996) Macromolecules 29: 6847
22. Hirai A, Horii F, Kitamaru R, Fatou JG, Bello A (1990) Macromolecules 23: 2913
23. Horii F, Hu S, Ito T, Odani H, Kitamaru R (1992) Polymer 33: 2299
24. Saito S, Moteki Y, Nakagawa M, Horii F, Kitamaru R (1990) Macromolecules 23: 3256
25. Nakaoki T, Hayashi H, Kitamaru R (1996) Polymer 37: 4833

26. Horii F, Hirai A, Kitamaru R (1987) Macromolecules 20: 2117
27. Horii F, Yamamoto H, Kitamaru R, Tanahashi M, Higuchi T (1987) Macromolecules 20: 2946
28. Horii F, Hirai A, Kitamaru R (1982) Polym Bull 8: 163
29. Horii F, Hirai A, Kitamaru R (1984) ACS Symp Series 260: 27
30. Horii F, Hirai A, Kitamaru R (1987) ACS Symp Series 340: 119
31. Horii F, Yamamoto H, Hirai A, Kitamaru R (1987) Carbohydr Res 160: 29
32. Strobl GR, Hagedorn W (1978) J Polym Sci Polym Phys Ed 16: 1181
33. Mutter R, Stille W, Strobl G (1993) J Polym Sci Part B: Polym Phys 31: 99
34. Mandelkern L (1985) Polym J 17: 337
35. Mandelkern L, Alamo RG (1993) In: Urban MW, Carver CD (eds) Structure-property relations in polymer. American Chemical Society, p 157
36. Yoon DY, Flory PJ (1977) Polymer 18: 509
37. Yoon DY, Flory PJ (1979) J Faraday Soc, Discuss Chem Soc 69: 288
38. Yoon DY (1978) J Appl Crystallogr 11: 531
39. Yoon DY, Flory PJ (1981) Polym Bull 4: 693
40. Vonk CG (1973) J Appl Crystallogr 6: 81
41. Santa Cruz C, Stribeck N, Zachmann HG, Balta Calleja FJ (1992) Macromolecules 24: 5980
42. Stribeck N, Alamo RG, Mandelkern L, Zachmann HG (1995) Macromolecules 28: 5029
43. With relation to more detail of this section, refer to standard text books, for example, Chapters 6 and 7 in "Nuclear magnetic resonance, principles and theory", Kitamaru R (1990) Elsevier Science Publishers Amsterdam
44. Instead of a 90° pulse, a 45° pulse is sometimes used to save time for the measurement. In this case, the equilibrium spectrum is obtainable by setting τ_ℓ as 3 times T_{1C}. For one pulse sequence only $1/\sqrt{2}$ times signal is obtainable but 5/3 times pulse sequences can be repeated for the same time as used for the measurement by the 90° pulse sequence. Since $(1/\sqrt{2}) \times (5/3)$ times signal is obtainable and the noise decreases by $\sqrt{3/5}$ times (the noise is thought to decrease proportionally to the reciprocal square root of the repetition number of the measurements), a spectrum with larger signal/noise ratios is obtainable in the same measuring time.
45. Wilson III CW, Pake GE (1953) J Polym Sci 10: 503
46. Fischer EW, Peterlin A (1964) Makromol Chem 74: 1
47. Olf HG, Peterlin A (1967) Kolloid-Z Z Polym 215: 97
48. Mandelkern L, Price JM, Gopalan M, Fatou JG (1966) J Polym Sci, Part A-2, 4: 385
49. Pranadi H, Manuel AJ (1980) Polymer 21: 303
50. VanderHart DL (1976) J Chem Phys 64: 830
51. Opella SJ, Waugh JS (1977) J Chem Phys 66: 4919
52. Earl WL, VanderHart DL (1979) Macromolecules 12: 762
53. Torchia DA (1978) J Magn Reson 30: 613
54. Axelson DE, Mandelkern L, Popli R, Mathieu P (1983) J Polym Sci Polym Phys Ed 21: 2319
55. Perez E, VanderHart DL, Crist B, Howard PR (1987) Macromolecules 20: 78
56. Voigt-Martin IG, Mandelkern L (1981) J Polym Sci Polym Phys Ed 19: 1769
57. Bassett DC, Hodge AM, Olley RH (1981) Proc R Soc London, A-No 377: 39
58. In this discussion, we have considered a molecular motion that includes two independent motions in order to explain qualitatively the fact that two noncrystalline phase involve the same T_{1C} and different T_{2C}'s. To analyze these phenomena more quantitatively, we have to evaluate T_{1C} and T_{2C} theoretically, assuming an adequate motional model. Refer to our article; Murayama K, Horii F, Kitamaru R (1983) Bull Inst Chem Res, Kyoto Univ, 61: 299
59. Till PH (1957) J Polym Sci 24: 301
60. Keller A (1957) Phil Mag 2: 1171
61. Fischer EW (1957) Z Naturforsch 12a: 753
62. Horii F, Kitamaru R (1978) J Polym Sci Polym Phys Ed 16: 265
63. Moller M, Gronski W, Cantow H-J, Hocker H (1984) J Am Chem Soc 106: 5093
64. Drotloff H, Emeis D, Waldron RF, Moller M (1987) Polymer 28: 1200

65 Ando I, Sorita T, Yamanobe T, Komoto T, Sato H, Deguchi K, Imanari M (1985) Polymer 26: 1864
66 Nakagawa M, Horii F, Kitamaru R, Tanaka Y, Sato H, Sakata Y, Hayase T (1987) Polym Prepr, Japan, 36: 3160
67 Axelson DE, Mandelkern L, Popli R, Mathieu P (1983) J Polym Sci Polym Phys Ed 21: 2319
68 Bassett DC, Khalifa BA, Olley RH (1977) J Polym Sci Polym Phys Ed 15: 995
69 Zhu Q, Horii F, Tsuji M, Kitamaru R (1989) J Soc Rheology, Japan, 17: 35
70 Smith P, Lemstra PJ (1980) J Material Sci 15: 505
71 Smith P, Lemstra PJ, Pijpers L, Kiel AM (1981) Coloid Polym Sci 259: 1070
72 Matsuo M (1985) J Soc Rheology, Japan, 13: 4
73 Flory PJ, Vrij A (1963) J Am Chem Soc 85: 3548
74 Gopalan M, Mandelkern L (1967) J Phys Chem 71: 3883
75 Pennings AJ, Zwijnenburg A (1979) J Polym Sci Polym Phys Ed 17: 1011
76 Matsuo M, Manley RStJ (1983) Macromolecules 16: 1500
77 Furuhata K, Yokokawa T, Miyasaka K (1984) J Polym Sci Polym Phys Ed 22: 133
78 Sumitani T, Zhu Q, Horii F, Odani M (1989) Polymer Preprints, Japan, 38: 4382
79 Kanemoto T, Tsuruta A, Tanaka K, Takeda M, Porter RS (1983) Polym J 15: 327
80 Randall JC (1975) J Polym Sci Polym Phys Ed 13: 1975
81 Hsieh ET, Randall JC (1982) Macromolecules 15: 353
82 Hsieh ET, Randall JC (1982) Macromolecules 15: 1402
83 Alamo RG, Chan EKM, Mandelkern L, Voigt-Martin IG (1992) Macromolecules 25: 6381
84 Alamo RG, Mandelkern L (1994) Thermochim Acta 238: 155
85 VanderHart DL, Perez E (1986) 19: 1902
86 Perez E, VanderHart DL (1987) Macromolecules 20: 78
87 Perez E, Bello A, Perena JM, Benvavente R, Aquilar MC (1989) Polymer 30: 1508
88 The weak resonance observed in the range 14–15 ppm can be assigned to the methyl groups of the ends in the main chain. This resonance, however, is not pertinent to our present analysis.
89 Imada K, Miyakawa T, Chatani Y, Tadokoro H, Murahashi S (1965) Makromol Chem 83: 113
90 Cesari M, Perego C, Mazzei A (1965) Makromol Chem 83: 196
91 Flory PJ (1953) Principles of polymer chemistry. Cornell Univ Pr, Ithaca, NY
92 Flory PJ, Yoon DY (1978) Nature (London) 272: 226
93 Bunn A, Cudby MEA, Harris RK, Packer KJ, Say BJ (1982) Polymer 23: 69
94 Lyerla JR, Yannoni CS (1983) IBM J Res Dev 27: 302
95 Gomez MA, Tanaka H, Tonelli AE (1987) Polymer 28: 2227
96 Ohkura M, Kanaya T, Kaji K (1992) Polymer 33: 3685
97 Kanaya T, Ohkura M, Kaji K, Furusaka M, Misawa M (1994) Macromolecules 27: 5609
98 Guenet J-M (1992) Thermoreversible gelation of polymers and biopolymers. Academic Press, London
99 Natta G, Corradini P, Peraldo M, Pegorano M, Zambelli A (1960) Rend Acd Naz Lincei 28: 539
100 Natta G, Peraldo M, Allegra A (1964) Makromol Chem 75: 215
101 Corradini P, Natta G, Ganis P, Temussi PA (1967) J Polym Sci, Part C, 16: 247
102 Lotz B, Lovinger AJ, Cais RE (1988) Macromolecules 21: 2375
103 Lovinger AJ, Lotz B, Davis DD (1990) Polymer 31: 2253
104 Lovinger AJ, Davis DD, Lotz B (1991) Macromolecules 24: 552
105 Lovinger AJ, Lotz B, Davis DD, Padden FJ (1993) Macromolecules 26: 3494
106 Lovinger AJ, Lotz B, Davis DD, Schumacher M (1994) Macromolecules 27: 6603
107 Rosa C, Corradini P (1993) Macromolecules 26: 5711
108 Chatani Y, Maruyama H, Noguchi K, Asanuma T, Shiomura T (1990) J Polym Sci, Part C, 28: 393
109 Chatani Y, Maruyama H, Asanuma T, Shiomura T (1991) J Polym Sci Polym Phys Ed 29: 1649
110 Tonelli AE, Schilling FC (1981) Acc chem Res 14: 233

111 Sozzani P, Galimberti M, Balbontin G (1992) Makromol Chem Rapid Commun 13: 305
112 Sozzani P, Simonutti R, Galimberti M (1993) Macromolecules 26: 5782
113 Aoki A, Hayashi T, Date T, Asakura T (1994) Polym Preprints, Japan, 43: 1473
114 Snyder RG, Schachtschneider JH (1964) Spectrochim Acta 20: 853
115 Schachtschneider JH, Snyder RG (1965) Spectrochim Acta 21: 1527
116 Tadokoro H, Kobayashi H, Kobayashi M, Yasufuku K, Mori M (1966) Rep Prog Polym Phys, Japan, 9: 181
117 Kobayashi M, Nakaoki T, Ishihara N (1990) Macromolecules 23: 78
118 Nakaoki T, Ohira Y, Hayashi H, Kitamaru R (1996) Proceedings of Society of Solid-State NMR for Materials (c/o Inst. Chem Res, Kyoto Univ), 20:94
119 Naylor CC, Meier RJ, Kip BJ, Williams KPJ, Mason SM, Conroy N, Gerrard DL (1995) Macromolecules 28: 2969

Editor: Prof. S. Okamura
Received: July 1997

Laser Light Scattering Characterization of Special Intractable Macromolecules in Solution

Chi Wu

Department of Chemistry, The Chinese University of Hong Kong, Shatin, N.T. Hong Kong, China
E-mail: chiwu@cuhk.edu.hk

This review summarizes the recent advances in characterization of some special intractable macromolecules in solution by laser light scattering. Since both static and dynamic laser light scattering (LLS) are theoretically well established, we focus the discussion on experimental details, such as the design of a high-temperature LLS spectrometer, the sample clarification, a novel differential refractometer, and some newly developed methods for data analysis which include a combination of static and dynamic LLS leading to molar mass distribution determination and LLS calibration of gel permeation chromatography.

List of Symbols and Abbreviations		104
1	Introduction	106
2	Basic Principles of Laser Light Scattering	107
2.1	Static Laser Light Scattering	108
2.2	Dynamic Laser Light Scattering	109
3	Experimental Section	111
3.1	Solvent Selection	111
3.2	High-Temperature Spectrometer	112
3.3	Solution Preparation at High Temperature	114
3.4	Differential Refractometer	117
4	Data Analysis	120
4.1	Conversion Between Translational Diffusion Coefficient Distribution and Molar Mass Distributions	120
4.2	Scaling of Translational Diffusion Coefficient D with Molar Mass M	121
4.2.1	Using a Set of Narrowly Distributed Standards	121
4.2.2	Using Two or More Broadly Distributed Samples	122
4.3	Combination of LLS with Other Off-Line Methods	123
4.3.1	With Intrinsic Viscosity	123
4.3.2	Gel Permeation Chromatography	124

5	Applications	126
5.1	Segmented Copolymer	127
5.2	A Polymer Mixture Containing Individual Chains and Clusters	129
5.3	Polymer Colloids	131
6	Conclusion	132
	References and Notes	132

List of Symbols and Abbreviations

Sect. 1

LLS	laser light scattering
GPC	gel permeation chromatography
FFF	field flow fractionation
$<M>$	average molecular weight
β	an integer number
M	molecular weight for monodisperse species
M_n	number-average molecular weight
M_w	weight-average molecular weight
M_z	intensity-average (z-average) molecular weight
$f_n(M)$	differential number distribution

Sect. 2

$I(t)$	scattering intensity at time t
$n(t)$	scattering photon counts at time t
$R_{vv}(\theta)$	excess Rayleigh ratio at angle θ
$<I>_{solution}$	time-averaged scattering intensity of solution
$<I>_{solvent}$	time-averaged scattering intensity of solvent
$<I>_{standard}$	time-averaged scattering intensity of standard
$<I>$	time-averaged scattering intensity
$R_{vv,standard}(\theta)$	Rayleigh ratio of standard at angle θ
n	refractive index of solution
$n_{standard}$	refractive index of standard
a	detection geometry parameter
$<R_g^2>^{1/2}$ or $<R_g>$	root-mean square z-average radius of gyration
q	scattering vector
K	optical constant
N_A	Avogadro number
dn/dC	specific refractive index increment
λ_o	wavelength of the laser light in vacuum
C	concentration of polymer solution
A_2	second virial coefficient
QELS	quasi-elastic light scattering

$G^{(2)}(t,q)$	intensity-intensity time correlation function at q and delay time t
t	time variable
t	delay time variable
PCS	photon correlation spectroscopy
$g^{(1)}(t,q)$	normalized first-order electric field time correlation function at q and delay time t
A	baseline
β	coherence parameter
I(t,q)	scattered intensity at time t and q
CONTIN	Laplace inversion program
G(Γ)	line-width distribution
Γ	line-width for monodisperse species
E(t,q)	scattered electric field at time t and q
<Γ>	average line-width
D	translational diffusion coefficient
k_d	diffusion second virial coefficient
f	dimensionless parameter
G(D)	translational diffusion coefficient distribution
<D>	average translational diffusion distribution
G	proportionality factor

Sect. 3

PPTA or Kevlar	poly(1,4-phenyleneterephthalamide)
Teflon and Tefzel	fluorocarbon polymers
PEL-PCL	poly(ethylene terephthalate-*co*-caprolactone)
Δn	refractive index difference between solvent and solution

Sect. 4

k_D	scaling constant in $D=k_D M^{-\alpha_D}$
α_D	scaling constant in $D=k_D M^{-\alpha_D}$
$f_w(M)$	differential weight distribution
$M_{w,calcd}^{DLS}$	weight-average molecular weight calculated on the basis of G(D) from DLS
$M_{w,measd}$	weight-average molecular weight measured from static LLS
ERROR(α_D)	error function of α_D
ERROR(k_D)	error function of k_D
[η]	intrinsic viscosity
k_η	scaling constant in $[\eta]=k_\eta M^{-\alpha_\eta}$
α_η	scaling constant in $[\eta]=k_\eta M^{-\alpha_\eta}$
$F_{w,cum}(M)$	cumulative weight distribution
SEC	size-exclusion chromatography
V	elution volume
A	calibration constant in $V=A+B \log(M)$
B	calibration constant in $V=A+B \log(M)$
C(V)	elution volume distribution

$M_{w,calcd}^{SEC}$ — weight-average molecular weight calculated on the basis of C(V) from GPC

Sect. 5

$M_{w,app}$	apparent weight-average molecular weight
ν	refractive index increment for the whole sample
$\nu(M)$	refractive index increment for copolymer with molecular weight M and weight distribution $f_w(M)$
$w_A(M)$	weight fraction of A for a given polymer chain with fixed M and $F_w(M)$
$w_B(M)$	weight fraction of B for a given polymer chain with fixed M and $F_w(M)$
ν_A	refractive index increment of polymer A
ν_B	refractive index increment of polymer B
w_A	weight fraction of A
w_B	weight fraction of B
ν_{s1}	ν in solvent 1
ν_{s2}	ν in solvent 2
$\nu_{A,s1}$	ν_A in solvent 1
$\nu_{B,s1}$	ν_B in solvent 1
$\nu_{A,s2}$	ν_A in solvent 2
$\nu_{B,s2}$	ν_B in solvent 2
$f_{w,app(M),s1}$	apparent weight distribution in solvent 1
$f_{w,app(M),s2}$	apparent weight distribution in solvent 2
$M_{w,L}$	low weight-average molecular weight polymer
$M_{w,H}$	high weight-average molecular weight polymer
X_L	weight fraction of low molecular weight polymer
X_H	weight fraction of high molecular weight polymer
$G_L(\Gamma)$	line-width distribution for low molecular weight polymer
$G_H(\Gamma)$	line-width distribution for high molecular weight polymer
A_L	area covered by $G_L(\Gamma)$
A_H	area covered by $G_H(\Gamma)$
A_r	area ratio A_L/A_H
R	radius of a colloidal particle
ϱ	density of particle
$(M_w)_{cal}$	calculated weight-average molecular weight
b	thickness of the hydrodynamic layer

1
Introduction

Differing from small molecules, typical polymers have an average molar mass of ~10^4 g/mol or higher and a wide distribution of chain length. For a given type of polymer, its properties, even its appearance, are greatly influenced not only by its average molecular weight, but also by its molecular weight distribution.

Therefore, the development and application of a polymer often require a precise characterization of these quantities.

A number of methods including laser light scattering (LLS) are routinely used to determine the average molecular weights and molecular weight distribution of a polymer. They include end-group chemical analysis, vapour pressure osmometry, membrane osmometry, sedimentation equilibrium, static (classic) LLS and very recently developed matrix-assisted time-fly mass spectroscopy as absolute methods in the sense that these do not require a calibration with a set of narrowly distributed polymer samples with known molecular weights. The relative methods include viscometry, gel permeation chromatography (GPC), field flow fractionation (FFF) and dynamic LLS.

The average molecular weight <M> of a polydisperse polymer sample is generally defined by

$$<M> = \int_0^\infty f_n(M) M^\beta \, dM \ / \ \int_0^\infty f_n(M) M^{\beta-1} \, dM \tag{1.1}$$

if β is an integer, where $f_n(M)$ is the number distribution of molecular weight M. Thus, $\beta=1$ for the number-average molecular weight (M_n), $\beta=2$ for the weight-average molecular weight (M_w) and $\beta=3$ for z-average molecular weight (M_z). For example, M_n is measured by end-group analysis and osmometry; M_w, by GPC, sedimentation equilibrium and static LLS; and M_z, by sedimentation equilibrium and dynamic LLS. In practice, the ratio of M_w/M_n is called the polydispersity index and is conveniently used to characterize the distribution width of a given polymer sample.

Static light scattering as a classic and absolute method has been long and widely used to characterize both synthetic and natural macromolecules [1–4]. In the last two decades, thanks to the advances in stable laser, ultra-fast electronics and personal computers, laser light scattering (LLS), especially dynamic LLS (denoted here by DLS), has gradually changed from a very special tool for physical chemists to a routine analytical tool in polymer laboratories or even to a daily quality-control device in production lines [5–8]. The LLS instruments commercially available nowadays are normally capable of making both static and dynamic measurements simultaneously.

2
Basic Principles of Laser Light Scattering

When a monochromatic, coherent light is incident into a dilute macromolecule solution, if solvent molecules and macromolecules have different refractive index, the incident light is scattered by each illuminated macromolecule to all directions [9, 10]. The scattered light waves from different macromolecules mutually interfere, or combine, at a distant, fast photomultiplier tube detector and produce a net scattering intensity $I(t)$ or photon counts $n(t)$ which is not uniform on the detection plane. If all macromolecules are stationary, the scattered light intensity at each direction would be a constant, i.e. independent of time.

However, in reality, all macromolecules in solution are undergoing constant Brownian motion, and this fact leads to fluctuation in I(t). The fluctuation rate can be related to the translational diffusion of the macromolecules. The faster the diffusion, the faster the fluctuation will be.

2.1
Static Laser Light Scattering

In static LLS, the angular dependence of the excess absolute time-averaged scattered intensity, known as the excess Rayleigh ratio, $R_{vv}(\theta)$, is normally measured, where $R_{vv}(\theta)=(<I>_{solution}-<I>_{solvent})/<I>_{standard} \cdot R_{vv,standard}(\theta) \cdot (n/n_{standard})^a$ with $<I>$ and n denoting the time-averaged scattering intensity and the refractive index, respectively, and $1 \leq a \leq 2$ depending on the detection geometry. Thus, if the scattering volume is selected by a slit, a=1, and if the scattering volume is selected by a pinhole much smaller than the beam diameter, a=2. If the solution is very dilute, $R_{vv}(\theta)$ at a relatively small scattering angle θ can be related to the weight-average molecular weight M_w, the second virial coefficient A_2 and the root-mean square z-average radius of gyration $<R_g^2>^{1/2}$ (or written as $<R_g>$ for simplicity) by the expression [9, 10]

$$\frac{KC}{R_{(vv)}(\theta)} \approx \frac{1}{M_w}\left(1 + \frac{1}{3}<R_g^2>q^2\right) + 2A_2C + \ldots \qquad (2.1)$$

where C is the mass concentration of the polymer component, $K=4\pi^2n^2 (dn/dC)^2/(N_A\lambda_o^4)$ and $q=(4\pi n/\lambda_o) \sin(\theta/2)$, with N_A, dn/dC, n and λ_o being the Avogadro constant, the specific refractive index increment, the solvent refractive index, and the wavelength of the laser light in vacuum, respectively. With $R_{vv}(\theta)$ measured at a series of C and θ, we can determine M_w, R_g, and A_2 by use of a Zimm plot, which allows both θ and C extrapolations to be made on a single grid [11]. Figure 1 shows a Zimm plot for an alternating copolymer of ethylene and tetraflu-

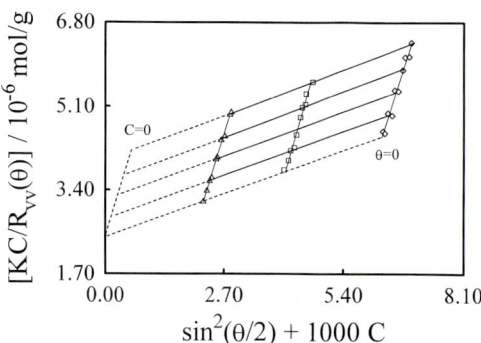

Fig. 1. Typical Zimm plot for an alternating copolymer of ethylene and tetrafluoroethylene ($M_w = 5.4 \times 10^5$ g/mol, $R_g = 45.4$ nm and $A_2 = 1.97 \times 10^{-4}$ mol mL/g^2) in diisobutyl adipate at 240 °C

oroethylene (M_w=5.4×10^5 g/mol, R_g=45.4 nm and A_2=1.97×10^{-4} mol mL/g^2) in diisobutyl adipate at 240 °C [12]. It should be noted that Eq. (2.1) is valid under the restriction that the polymer solution exhibits no adsorption, no fluorescence, and no depolarized scattering. As for rigid and nearly rigid rods causing depolarized scattering, readers can refer to the excellent review article of Russo and the references therein [13]. As for the correction of adsorption and fluorescence, readers are advised to refer to the characterization of Kevel in concentrated sulfuric acid by Qing et al. [14–16].

2.2
Dynamic Laser Light Scattering

DLS measures the intensity fluctuation instead of the average light intensity (this is where the word 'dynamic' comes from), and its essence may be explained as follows. When the incident light is scattered by a moving macromolecule or particle, the detected frequency of the scattered light will be slightly higher or lower than that of the original incident light owing to the Doppler effect, depending on whether the scatterer moves towards or away from the detector. Thus, the frequency distribution of the scattered light is slightly broader than that of the incident light. This explains why dynamic light scattering is sometimes called quasi-elastic light scattering (QELS). The frequency broadening (~10^5–10^7 Hz) is so small in comparison with the incident light frequency (~10^{15} Hz) that, if not impossible, it is very difficult to detect it. However, it can be effectively recorded in the time domain through an intensity-intensity time correlation function $G^{(2)}(t,q)$ in the self-beating mode. Thus, dynamic light scattering is also known as photon correlation spectroscopy (PCS).

$G^{(2)}(t,q)$ can be related to the normalized first-order electric field time correlation function $|g^{(1)}(t,q)|$ by [9, 10]

$$G^{(2)}(t,q) = <I(t,q)I(0,q)> = A[1+\beta|g^{(1)}(t,q)|^2] \qquad (2.2)$$

where A is the measured base line, β is a parameter depending on the coherence of the detection optics, and t is the delay time.

For a monodisperse sample, $g^{(1)}(t,q)$ is theoretically represented by

$$g^{(1)}(t,q) = Ge^{-\Gamma t} \qquad (2.3)$$

where G and Γ are the proportionality factor and the line-width respectively. For dilute solutions, Γ measured at a finite scattering angle is related to C and q by [17]

$$\Gamma = q^2 D(1+k_d C)(1+f<R_g^2>_z q^2) \qquad (2.4)$$

Here D is the translational diffusion coefficient of the solute molecule at $C \to 0$ with C the mass concentration of the solute, k_d the diffusion second virial coefficient, f a dimensionless parameter depending on polymer chain structure and solvent, and $<R_g^2>$ the mean square radius of gyration of the polymer chain. Hence, for C and q small enough, Eq. (2.3) may be approximated by

$$g^{(1)}(t,q) = G_e^{-q^2 Dt} \tag{2.5}$$

The proportionality factor G may depend on many characteristics of the solute polymer, but since D is the only polymer-dependent variable in Eq. (2.5), it is reasonable here to treat G as a function of D only for a homologues series of polymers. Thus, for a polydisperse polymer sample with a continuous distribution of molecular weight M, Eq. (2.5) may be generalized as

$$g^{(1)}(t,q) = \int_0^\infty G(D) e^{-q^2 Dt} dD \tag{2.6}$$

where G(D) is called the translational diffusion coefficient distribution. This equation is the basic of the entire discussion in the present paper. Note that since $g^{(1)}(t,q) \to$ unity as t goes to zero, we have

$$\int_0^\infty G(D) dD = 1 \tag{2.7}$$

that is, G(D) is normalized. Equation (2.6) indicates that once $g^{(1)}(t,q)$ is determined from $G^{(2)}(t,q)$ through Eq. (2.2), G(D) can be computed by Laplace inversion [18–24]. Among many others, the program called CONTIN [24] is one of the most widely used and accepted for this computation. However, due to the bandwidth limitation of photon correlation instruments, some unavoidable noises, and a limited number of data points, the data obtained for $g^{(1)}(t,q)$ do not always provide information necessary and sufficient to determine G(Γ) uniquely. This difficulty is well-known as the ill-posed problem. Thus, in practice, reducing the noises in the measured $G^{(2)}(t,q)$ becomes more important than choosing a program for data analysis.

Figure 2 illustrates the $g^{(1)}(t,q)$ data for chitosan ($M_w=1.06\times10^5$ g/mol and $<D>=5.92\times10^{-8}$ cm^2/s) in aqueous 0.2 M CH$_3$COOH/0.1 M CH$_3$COONa at 25 °C, θ=45° and C=4.96×10^{-4} g/ml. Here, $<D>$ is the average diffusion coefficient given by

$$<D> = \int_0^\infty G(D) D \, dD \tag{2.8}$$

Figure 3 shows the G(D) function for the same system as in Fig. 2 at $\theta \to 0$ and $C \to 0$.

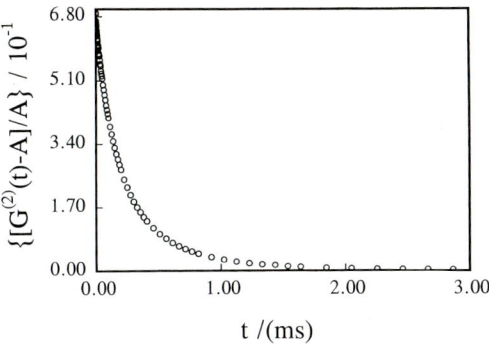

Fig. 2. Typical normalized intensity-intensity time correlation function for chitosan ($M_w = 1.06 \times 10^5$ g/mol and $<\Gamma> = 2.19$ ms) in 0.2 M CH_3COOH / 0.1 M CH_3COONa aqueous solution at $T = 25$ °C, $\theta = 45°$ and $C = 4.96 \times 10^{-4}$ g/mol

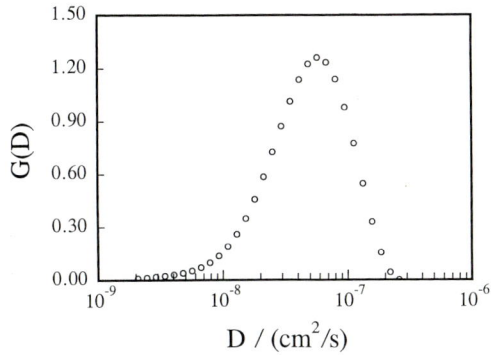

Fig. 3. Typical translational diffusion coefficient G(D) for chitosan ($M_w = 1.06 \times 10^5$ g/mol and $<D> = 5.92 \times 10^{-8}$ cm²/s) in 0.2 M CH_3COOH / 0.1 M CH_3COONa aqueous solution at $T = 25$ °C, $\theta \to 0$ and $C \to 0$

3
Experimental Section

3.1
Solvent Selection

If a macromolecule can be dissolved in more than one solvent, the choice of the solvent for laser light scattering measurement should be made generally according to the following three criteria: 1) it is colorless so that the adsorption correction can be avoided, 2) it has a higher contrast, i.e. a higher value of the specific refractive index increment dn/dC, so that the signal-to-noise ratio is increased, and 3) it is less polar and less viscous so that the solution may be clarified more easily.

Sometimes in practice, we may have no choice of solvent for a given polymer. For example, poly(1,4-phenyleneterephthalamide) (PPTA or Kevlar) is only soluble in very strong acids which are viscous. In such cases, ultracentrifugation instead of filtration has to be used to remove dust particles from the solution [14–16, 25]. As for copolymers, the selection of proper solvents is even more difficult, because at least two solvents which satisfy the above mentioned three criteria are needed. For this reason, reported characterization of copolymers is quite limited [26–29].

One of the challenging problems in the application of LLS methodology is to study static and dynamic properties of fluorocarbon polymers, such as Teflon and Tefzel (registered trademarks of Du Pont), which defy the characterization owing to their insolubility in ordinary solvents. It is this unique solubility that makes these fluorocarbon polymers very useful in many applications, but greatly annoys those who wish to explore their solution properties. However, it was found that some solvents were capable of dissolving them at high temperature, but the finding was not enough to solve all characterization problems because the technique for clarification and measurement of the solutions at high temperature remained to be established.

3.2
High-Temperature Spectrometer

In order to characterize a polymer soluble only at high temperature, many difficulties had to be overcome before light-scattering measurement was actually made. Thus, a special high-temperature LLS spectrometer was first designed and developed at State University of New York in Stony Brook, and then its technique was transferred to both Du Pont and BASF [30–35].

Figure 4 shows a schematic diagram of the high-temperature LLS spectrometer at Du Pont. In it, a thermally-controllable plate (13) used as a heat sink isolates the rotary table (12) from the outer thermostat (3) by means of two sets of stainless steel standoffs (14). The outer brass thermostat (3) is isolated from the room with 0.5-in.-thick porous silicone rubber. This arrangement creates an oven that allows the temperature to keep easily in the 320–340 °C range with temperature fluctuations of less than 0.2 °C. A glass (vacuum) jacket of 2.25-in. o.d. isolates the inner thermostat from the oven. The vacuum jacket reduces the temperature gradient in the light scattering cell. The inner thermostat has a separate temperature controller and a miniature platinum resistance thermometer that can be connected to a digital voltmeter through the vacuum jacket by means of ceramic feed throughs. With this design, short-term (20 min) control of ±0.05 °C, intermediate-term (60 min) control of ±0.1 °C, and long-term control of ±0.5 °C can be achieved at 340 °C even in the absence of a vacuum. Long-term temperature stability depends partially on room-temperature fluctuations even in the presence of the outer thermostat and the isolation

Laser Light Scattering Characterization of Special Intractable Macromolecules in Solution

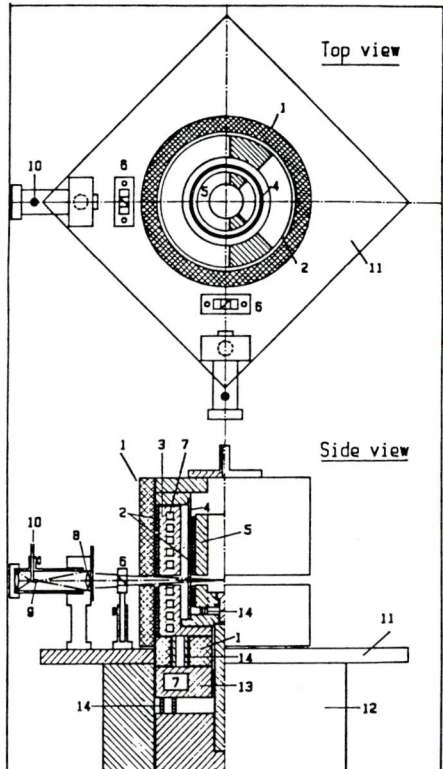

Fig. 4. Schematic top and side views of the high-temperature thermostat and detection system of the light scattering spectrometer: (1) silicon rubber insulation; (2) heating wires for the brass thermostats; (3) outer brass thermostat with fluid circulation facilities; (4) vacuum glass jacket for thermal isolation made of precision polished glass of 2.25-in. o.d. with Kovar seals at both ends of a stainless-steel temperature shield with precision polished glass of 2.25-in. o.d.; (5) inner brass thermostat, which has a separate temperature controller and a thermometer and can accommodate a light scattering cell up to 27-mm o.d; (6) Glan-thompson polarizers; (7) fluid circulation paths; (8) lens; (9) field aperture; (10) optical fiber bundle; (11) rotating plate for multiple detectors; (12) RT-200 Klinger rotary table with 0.01° step size; (13) cooling plate to isolate the outer thermostat from the rotary table; and (14) stainless steel standoffs for thermal isolation

between the two thermostats. Figure 5 shows typical temperature fluctuations of the inner thermostat at 340 °C. To mention more, a high-temperature LLS detector coupled with GPC has recently been developed and the determination of the molar mass distribution of poly(phenylene sulfide) in 1-chloronaphthalene at 220 °C has been made possible with it [36]. The advantage of on-line coupling LLS with GPC is obvious, since GPC is a fractionation method and LLS allows an absolute molar mass measurement and hence makes the calibration of GPC columns.

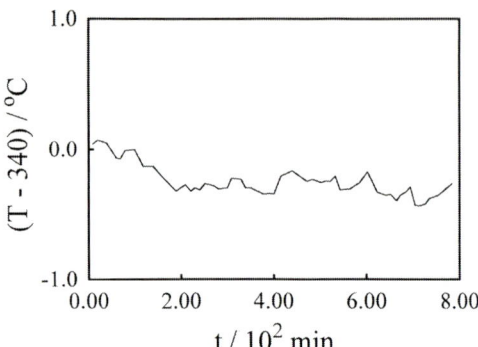

Fig. 5. Typical temperature fluctuations in the inner thermostat at 340 °C. Intermediate-term (1 h) temperature fluctuations were ± 0.1 °C. Long-term (10 h) temperature fluctuations were ± 0.5 °C

3.3
Solution Preparation at High Temperature

In order to prepare and clarify a polymer solution at a temperature higher than 200 °C, two different apparatuses were developed [30, 33]. Importantly, they are able (1) to dissolve a polymer under an inert atmosphere without losing solvent and without building up inner pressure due to solvent evaporation, and (2) to transfer the solution into a filtration device by a remote control.

Figure 6 shows a specially designed dissolution/filtration apparatus which can be placed inside a small oven. Known weights of a polymer sample and a filtered solvent, as well as a small glass-enclosed magnetic stirrer, are placed in A at room temperature. The solution vessel (A) is then connected to the precleaned filter (B). After degassing followed by introduction of nitrogen, both stopcocks are closed and the oven is heated to the desired temperature to dissolve the polymer while the solution is stirred. When the polymer is considered to have been completely dissolved, the solution vessel is turned 180° by means of the seal glass joint (J). This allows the polymer solution to be transferred from A into B without exposure to air. A gentle pressure using nitrogen is applied to force the polymer solution to pass through the fine-grade sintered glass filter (f) and to move directly into the precleaned dust-free cylindrical light scattering cell (C). In this way, dust-free polymer solutions can be successfully prepared, keeping the temperature high.

Figure 7 shows another type of dissolution/filtration apparatus. Sleeve A (with no bottom) is joined to the shaded stopper, which is connected to a reflux condenser by means of a greaseless glass joint. Cup (B), with a magnetic stirrer (E) sitting on top of the fine-grade sintered glass filter (F_2), allows solution clarification. First, a dust-free solvent and a polymer are placed in B of the argon-filled apparatus which has the preattached dust-free light-scattering cell (D). The reflux condenser flushed with argon is then inserted taking care that the apparatus with the polymer and solvent is under an inert atmosphere at room tem-

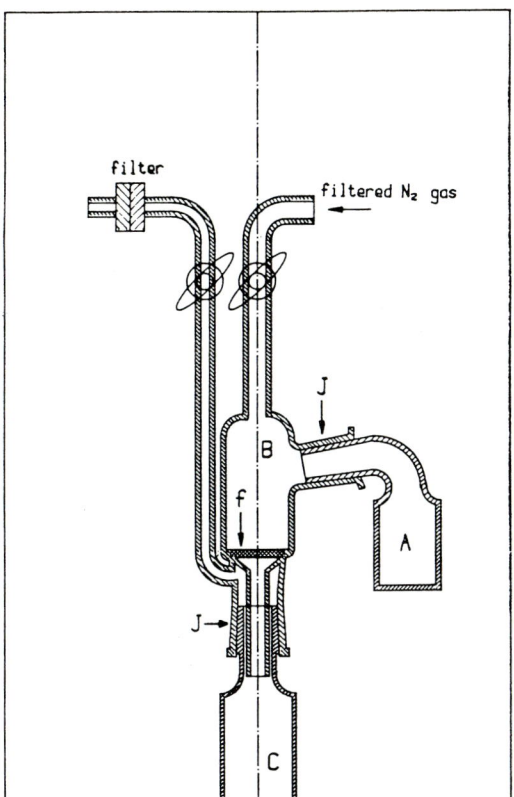

Fig. 6. High-temperature dissolution/filtration apparatus. The entire apparatus is placed in a high-temperature oven controlled at 250 ± 2 °C. (A) Solution vessel, where known weights of polymer and solvent as well as a small glass-enclosed magnetic stirrer are introduced. (B) a fine-grade sintered glass filter, connected to A and C by means of clean seal glass joints (J) (14/20, Wheaton Scientific). (f) Fine-grade sintered glass filter. (C) Cylindrical light scattering cell of 27-mm o.d. with a clean seal glass joint

perature. The entire apparatus is set in a small oven and the temperature is raised to the desired value. When the polymer is completely dissolved, an argon pressure is applied through the sintered glass filter (F_1) to let the polymer solution move from B to C. The additional pressure difference plus gravity will filter the polymer solution directly into the dust-free light-scattering cell (D). After the filtration process is completed, the additional argon pressure is released. The stopcock above the upper filter is closed during a light-scattering experiment. Briefly, extreme care must be taken when the test solution is prepared and subjected to light scattering measurements at temperatures near the boiling point of the solvent because the pressure build-up could cause an explosion. This points to the absolute necessity of having a pressure releasing mechanism.

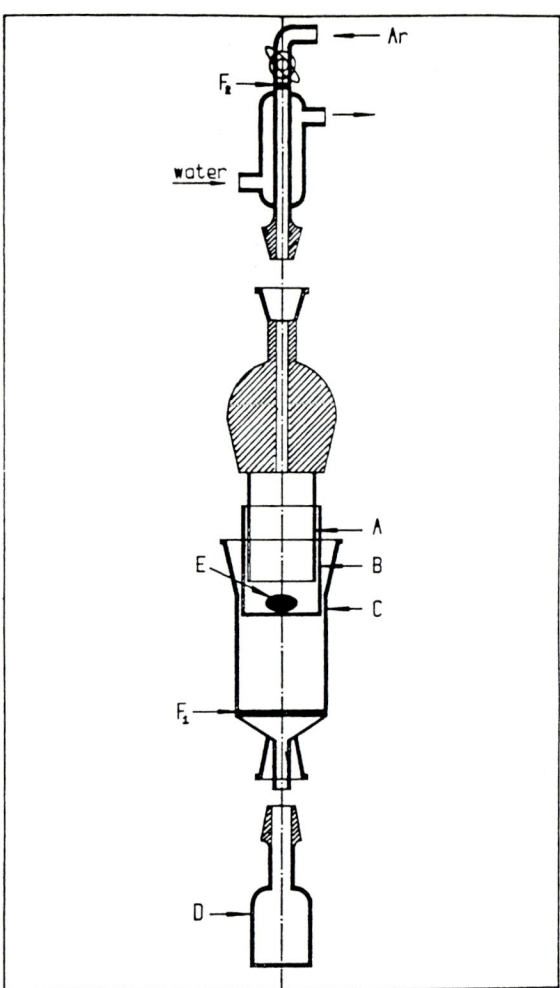

Fig. 7. Separate components of a high-temperature dissolution/filtration apparatus. The assembled apparatus, as shown in Fig. 1 of Ref. 30, is placed in a high-temperature oven. (A) Cylindrical insert (with on bottom) with a diameter ~2 mm less than that of the solution vessel (B); (C) Filtration section with a fine-grade sintered glass filter (F_1) and ground glass joints to a light scattering cell (D) and a ground glass joint adapter for the water cooled condenser which is located outside the temperature-controlled oven. (E) Magnetic stirring bar. The shaded area denotes volume reduction so that the volume accessible by vapour phase is no more than a few times the fluid phase. The miniature water-cooled condenser has a coarse-grade sintered-glass filter (F_2) so that the entire system is always isolated from external dust. The greaseless stopcock above F_2 is for operating the apparatus as a closed system, for introduction of low vacuum in order to degas the solvent before dissolution, for filling the apparatus with inert gas, such as argon, in order to alleviate chemical decomposition, and for releasing a possible pressure build-up at high temperature if chemical decomposition takes place. The entire apparatus is portable and can be inserted into the high-temperature light-scattering spectrometer with the light-scattering cell (D) and part of the filtration component (C) controlled at a given high temperature

3.4
Differential Refractometer

One of the most important parameters in static LLS is the specific refractive index increment (dn/dC), defined as $lim_{C \to 0}(\partial n/\partial C)_{T,P,\lambda}$. Since this parameter is not an intrinsic property of the polymer, the conditions of fixing temperature T, pressure P and wavelength of light in vacuum λ are needed in its definition. Note that, according to Eq. (2.1), an error of E% in dn/dC will lead to an error of 2E% in the derived M_w.

The refractive index increment Dn is normally measured by using either a differential refractometer or an interferometer. In the former, light is refracted at the boundary between the sample and a reference liquid. Commonly, the beam displacement is measured directly and then converted to Δn by multiplying by a calibrated constant which can be obtained by using a solution with an accurately known refractive index difference Δn [37]. This is not an absolute method since the constant has to be calibrated under the same conditions as in the light scattering measurements. In the latter, two light beams with identical geometrical paths traverse two different optical paths. One passes through the sample and the other through the reference liquid. This method relies on the interference of the two beams. Its details can be found elsewhere [38, 39]. In a high-temperature LLS measurement, a deformed cylindrical light scattering cell is preferred to the conventional divided differential refractometer cuvette in which the exit laser beam is refracted by the solution/air interface [30].

Figure 8a shows the design of a recently developed and commercially available refractometer (ALV GmbH, Langen, Germany). A small pinhole (P) with a diameter of 400 µm is illuminated with a laser light. The illuminated pinhole is imaged to a position-sensitive detector (PD) (Hamamatsu S 3932) by a lens (L) located at an equal distance from the pinhole and the detector, where the distance is four times the focal-length (f=100 mm) of the lens. Thus, this novel refractometer uses a (2f-2f) design instead of a conventional (1f) design which uses parallel incident light beams and makes the distance between the detector and the lens equal to only one focal length. A temperature-controlled refractometer cuvette (C) (Hellma 590.049-QS) is placed just in front of the lens. It is a flow cell and has a volume of ~20 mL, which is divided by a glass plate at ~45° into two compartments. The pinhole, the cuvette, the lens and the detector are rigidly mounted on a small optical rail. The refractometer has dimensions of only 40 cm in length, 15 cm in width and 10 cm in height, and the length can be easily reduced to 20 cm with another lens if necessary. The output voltage (−10 to 10 volts) from the position-sensitive detector is proportional to the displacement of the light spot from the center of the detector, and can be measured by a digital voltmeter or an analog-to-digital data acquisition system and a personal computer.

Figure 8b shows the basic principle and the light path of the refractometer, where θ', θ", θ''', and the cuvette are drawn enlarged to make the details clear. If both the compartments are filled with a solvent (i.e. n=n_o), the illuminated pinhole will be imaged at point O. However, if the solvent in one of the compartments is replaced by a dilute polymer solution with a slightly different refractive index (i.e. n=n_o+Δn), the light will be bent first by the glass plate, then by the

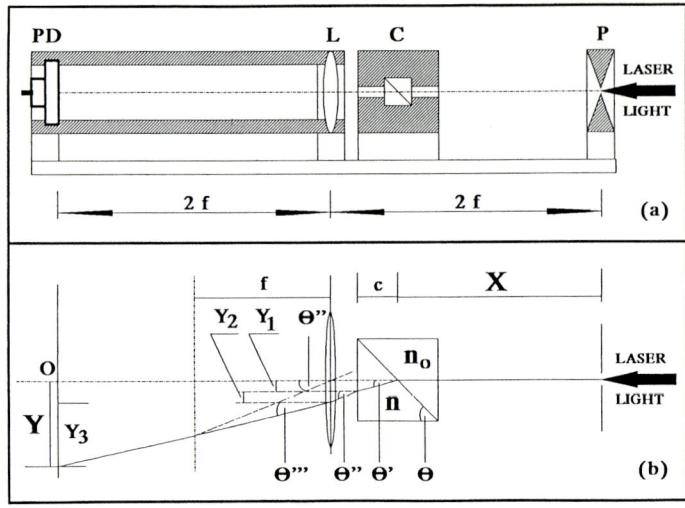

Fig. 8 a. Schematic view of a novel differential refractometer (produced by ALV GmbH, Langen, Germany), which consists of a pinhole (P), a differential refractometer cuvette (C), a lens (L, f = 10 cm) and a position sensitive detector (PD). All components are rigidly mounted on a 40 cm long optical rail. **b** Light path in which one compartment of the cuvette contains a solvent with refractive index n and the other contains a solution with slightly different refractive index $n = n_o + \Delta n$. The cuvette and angles θ', θ'' and θ''' (actually very small, ~ 0.01 radian) are enlarged to make the light path distinct

cuvette wall and finally by the lens. The image is shifted away from the point o by a distance of Y. Figure 7b shows that

$$Y = Y_1 + Y_2 + Y_3 = c\tan(\theta') + (2f - X - c)\tan(\theta'') + 2f\tan(\theta''') \quad (3.1)$$

and

$$f\tan(\theta'') = f\tan(\theta''') + c\tan(\theta') + (2f - X - c)\tan(\theta'') \quad (3.2)$$

where c, X and θ are constants. Snell's law gives

$$n_o\sin(90-\theta) = (n_o + \Delta n)\sin(90 - \theta - \theta') \quad (3.3)$$

and

$$(n_o + \Delta n)\sin(\theta') = \sin(\theta'') \quad (3.4)$$

where θ', θ'' and θ''' are actually so small because Δn is in the order of 10^{-4} RI units that we may set $\sin(\theta') = \theta'$, $\sin(\theta'') = \theta''$, $\tan(\theta') = \theta'$, $\tan(\theta'') = \theta''$ and $\tan(\theta''') = \theta'''$. Combination of Eqs (3.1)–(3.4) leads to

$$Y = K\Delta n \quad (3.5)$$

where $K = [X + c''(1 - 1/n_o)]\tan(90° - \theta)$. For a given optical set-up and solvent, X, c, θ, n_o, and hence K are constants. Equation (3.5) shows that the signal is proportional to Δn, and the larger the value of X, the higher the sensitivity ($Y/\Delta n$) is.

This means that the cuvette should be placed as close as possible to the lens in the experimental set-up.

In the (2f-2f) design, the detector and the pinhole (acting as a light source) are placed at the exact imagining positions along the optical axis of the lens. This configuration is optically equivalent to placing the detector directly behind the pinhole, so that the laser beam drift is eliminated. In comparison with the conventional differential refractometer, this novel design has made the measurement of Δn much easier and provides reliable and accurate values for dn/dC since it is stable and all the results can be recorded and averaged on a computer. Figure 9 shows the concentration dependence of Δn for a 13% PET-PCL copolymer in three different solvents. The lines represent the least-square fits to the data points.

The refractometer with its present dimensions can be easily installed into any existing laser light-scattering spectrometer together with the laser source, the thermostat and the computer, as exemplified in Fig. 10. The optical glass plate

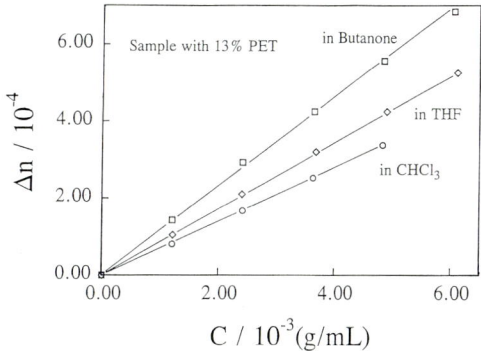

Fig. 9. Possible arrangement of the novel differential refractometer with an existing laser light scattering spectrometer

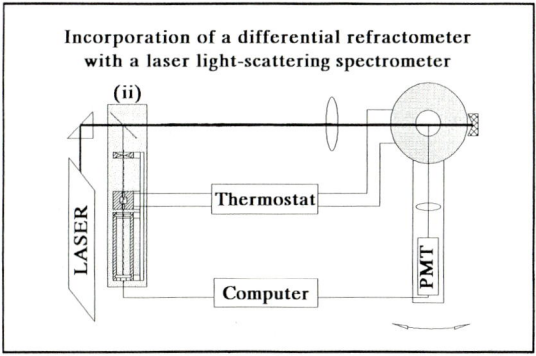

Fig. 10. Concentration dependence of the refractive index difference (Δn) between the polymer solution and solvent for a 13% PET-PCL copolymer. The lines represent the least-square fits to the measured data

placed in the laser light path at 45° reflects laser light by about 4%, and the reflected light is used as the light source. With this design, the light scattering and the refractive index increment can be simultaneously measured under the identical experimental conditions of wavelength and temperature. The details of this novel spectrometer have been described elsewhere [40].

4
Data Analysis

The methods of analyzing data for the concentration and angular dependence of the time-average scattering light intensity and the intensity-intensity time correlation function can be found in many LLS books and related literature. In this section, we will mainly concern ourselves with how to combine static and dynamic LLS results to characterize special polymers in regard not only to the average molar mass, but also to the molar mass and composition distributions.

4.1
Conversion Between Translational Diffusion Coefficient Distribution and Molar Mass Distributions

Though not involving fractionation, DLS allows estimation of the molecular weight distribution of a polymer. The principle is as follows. For a polydisperse polymer consisting of n homologous species, it is well known that $G(D_i)$ for species i at vanishingly small C and q is given by

$$G(D_i) = M_i w_i / \sum_{j=1}^{n} M_j w_j \quad (4.1)$$

where w_i denotes the weight fraction of species i with molecular weight M_i. For a continuous distribution of molecular weight this gives

$$G(D)dD = M f_w(M) dM / M_w \quad (4.2)$$

where $f_w(M)$ denotes the weight distribution of molecular weight M. Thus, we get

$$f_w(M) = (M_w / M) G(D) (dD/dM) \quad (4.3)$$

Empirically we have for a series of homologous polymers [41]

$$D = k_D M^{-\alpha_D} \quad (4.4)$$

where k_D and α_D are constants. Experimentally, for a flexible polymer, $0.5 < \alpha_D < 0.6$ in a good solvent and $\alpha_D = 0.5$ in a Flory Θ solvent; for a rigid rod-

Laser Light Scattering Characterization of Special Intractable Macromolecules in Solution 121

like chain, $\alpha_D=1$; and for a semi-rigid worm-like chain, $0.6<\alpha_D<1$. With Eq. (4.4), Eq. (4.3) is transformed to

$$f_w(M) = k_D M_w M^{-\alpha_D-2} G(D) \qquad (4.5)$$

which indicates that $f_w(M)$ can be determined if $G(D)$ is obtained by Laplace inversion of Eq. (2.6) and the values of the parameters k_D and α_D are available from separate sources. This is the basic idea of the method which allows information about $f_w(M)$ to be derived by DLS.

Since, as noted above, the success in determining $G(D)$ is actually not in the choice of a computer program for Laplace inversion but reducing the noise level in measured $g^{(1)}(t,q)$. Thus, it is crucial that the solution is cleaned (i.e. "dust-freed") very thoroughly before it is subjected to laser light scattering measurements. For example, in studies conducted by the author, efforts were made to ensure that the relative difference between the measured and calculated baselines did not exceed 0.1%. The error analysis related to the above problem can be found elsewhere [42, 43].

4.2
Scaling of Translational Diffusion Coefficient D with Molar Mass M

4.2.1
Using a Set of Narrowly Distributed Standards

The most straightforward method for calibrating the relationship between D and M is to measure both D and M for a set of monodisperse samples with different molecular weights. In reality, the monodisperse samples have to be replaced by narrowly distributed standards made available either by relevant living polymerization or by fractionation of a broadly distributed sample. However, only a few kinds of polymers, e.g. polystyrene and poly(methyl methylacrylate), can actually be prepared so as to have a sufficiently narrow molecular weight distribution ($M_w/M_n \sim 1.1$), and the fractionation is very time consuming. Thus, the straightforward calibration of the D vs M relation is not always practical.

Figure 11 shows the plot of log(D) versus log (M) for a set of narrowly distributed polystyrene standards in toluene at 20 °C [44]. The line represents a least-square fitting to the data points, giving $D(cm/s)=3.64\times10^{-4}M^{-0.577}$. Using this relation, we were able to estimate the molar mass distribution of polystyrene by using only DLS [45]. However, as noted above, in reality, only a few kinds of polymers can be prepared to have narrow enough distributions of molecular weight. Hence, we often have to satisfy ourselves with more broadly distributed samples having different average molecular weights. This means that special analytical methods have to be developed to calibrate or scale the translational diffusion coefficient D and molar mass M by using from broadly distributed samples. Ideas on this theme are described in the following sections.

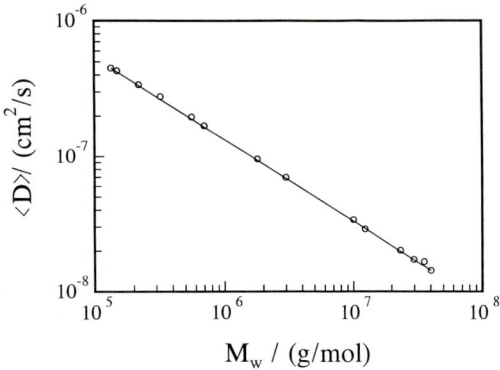

Fig. 11. Double logarithmic plot D vs M for polystyrene in toluene at 20 °C. The line represents a least-square fits to the data points, giving $D(cm/s) = 3.64 \times 10^{-4} M^{-0.577}$ (Ref. 42)

4.2.2
Using Two or More Broadly Distributed Samples

As can be easily shown, it follows from Eqs.(4.4) and (4.5) that

$$M_w = (k_D)^{1/\alpha_D} / \int_0^\infty G(D) D^{1/\alpha_D} dD \qquad (4.6)$$

The quantity on the right-hand side is denoted by $M_{w,calc}^{DLS}$, i.e.

$$M_{w,calc}^{DLS} = (k_D)^{1/\alpha_D} / \int_0^\infty G(D) D^{1/\alpha_D} dD \qquad (4.7)$$

For any given set of k_D and α_D, it can be calculated when $G(D)$ is determined from DLS measurements. We denote it for a polydisperse sample i by $(M_{w,calc}^{DLS})_i$. Then we get

$$\frac{(M_{w,calc}^{DLS})_i}{(M_{w,calc}^{DLS})_j} = \left[\int_0^\infty G_j(D) D^{1/\alpha_D} dD\right] / \left[\int_0^\infty G_i(D) D^{1/\alpha_D} dD\right] \qquad (4.8)$$

where $G(D)_i$ is for sample i. The right-hand side can be calculated from measured $G(D)_i$ and $G(D)_j$ for any α_D, and if the chosen value of α_D happens to be equal to that for the system under study, the resulting value of $(M_{w,calcd}^{DLS})_i / (M_{w,calcd}^{DLS})_j$ should agree with the value of $(M_w)_i / (M_w)_j$ which can be determined by SLS. In reality, the desired α_D will be reached by using a computer program which seeks a minimum of ERROR(α_D) defined by

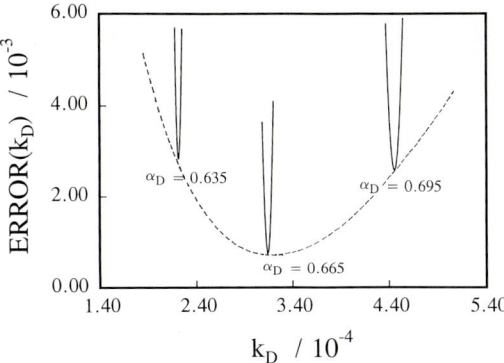

Fig. 12. Plot of ERROR(k_D) calculated with data for five Chitosan samples with different weight average molecular weights. Here the minimum of the dashed curve corresponds to $\alpha_D = 0.665 \pm 0.015$ and $k_D = (3.14 \pm 0.20) \times 10^{-4}$

$$\text{ERROR}(\alpha_D) = \sum_i^N \sum_j^N [\frac{(M_w)_i}{(M_w)_j} - \frac{(M_{w,calc}^{DLS})_i}{(M_{w,calc}^{DLS})_j}]^2 \quad (4.9)$$

where N is the total number of the polydisperse samples examined.

Next, with the α_D value so determined, we compute $(M_{w,calcd}^{DLS})_i$ for each of the N samples from Eq. (4.7) by varying k_D and seek a k_D value which minimizes ERROR(k_D) defined by

$$\text{ERROR}(k_D) = \sum_i^N [\frac{(M_w)_i}{(M_{w,calc}^{DLS})_i} - 1]^2 \quad (4.10)$$

Figure 12 depicts a plot of ERROR(k_D) at three values of α_D calculated from SLS and DLS data for five chitosan samples with different M_w. It is seen that for each chosen α_D ERROR(k_D) shows a sharp minimum, but the location and height of the minimum varies significantly with α_D and the minimum becomes the smallest at $\alpha_D=0.665$ and $k_D=3.14\times10^{-4}$. With these values, the molecular weight distributions of chitosan samples have been successfully characterized [18].

4.3
Combination of LLS with Other Off-line Methods

4.3.1
Intrinsic Viscosity

If only one broadly distributed sample is available, we have to resort to another method to determine the relation between D and M. One of them is to estimate α_D from the Mark-Houwink equation for intrinsic viscosity. It is known that the intrinsic viscosity [η] can be scaled with M by the Mark-Houwink equation, i.e.

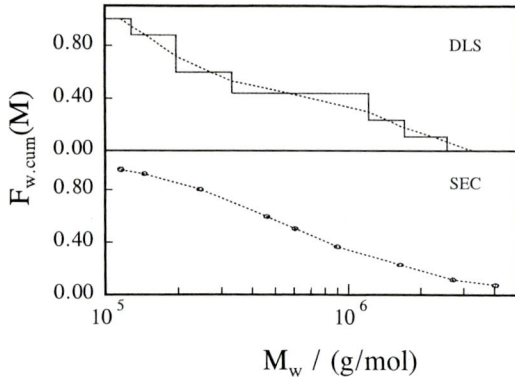

Fig. 13. Comparison of cumulative weight distributions $F_{w,cum}(M)$ [$= \int_M^\infty f_w(M)\,dM$] obtained by DLS and SEC (size exclusion chromatography) for a linear polyethylene in 1,2,4-trichlorobenzene at 135 °C

$[\eta]=k_\eta M^{\alpha_\eta}$, and according to Flory and also to de Gennes [41, 46], α_D may be equal to$=(\alpha_\eta+1)/3$. With α_D estimated from α_η by this relation, M_w from static LLS and G(D) from dynamic LLS can be used to determine k_D as described above. Chu et al. [47,48] successfully applied this method to linear polyethylene in 1,2,4-trichlorobenzene at 135 °C for which α_D was estimated from the reported value of $\alpha_\eta=0.72$. [49]

Figure 13 shows the resulting cumulative weight distribution $F_{w,cum}(M)$ [$=\int_M^\infty f_w(M)\,dM$] and compares it with the result obtained by high-temperature size exclusion chromatography (SEC). The agreement is reasonably good, but it should be noted that the weight distribution from LLS is usually narrower and more skewed toward the high molar mass than that from GPC because the scattered light intensity is proportional to the square of molar mass so that higher molar mass species weigh more in LLS.

4.3.2
Gel Permeation Chromatography

The static laser light scattering apparatus used as an on-line GPC detector has been popular for a while. Here, we illustrate another but less known method of combining the results from (gel permeation chromatography) and DLS. The basic principle is as follows: There is a similarity between these two tools in that the translational diffusion coefficient D obtained by DLS and the elution volume V in GPC are related to the hydrodynamic size of a given macromolecule. In a first approximation, if the hydrodynamic size is proportional to the molar mass, we have

$$V = A + B \log(M) \qquad (4.11)$$

where A and B are constants similar to k_D and α_D. It should be noted that this approximation simplifies but does not affect the following discussion. Combining of Eqs. (4.4) and (4.11) leads to

$$V = A + B \log(D) \tag{4.12}$$

where $A = A + B \log(k_D)/\alpha_D$ and $B = -B/\alpha_D$. Furthermore, we get from Eq. (4.12)

$$V^2 = A^2 + 2AB \log(D) + B^2 \log^2(D) \tag{4.13}$$

Averaging both sides of Eqs. (4.12) and (4.13) over the concentration profile C(V) in GPC, we obtain

$$<V> = A + B <\log(D)> \tag{4.14}$$

and

$$<V^2> = A^2 + 2AB <\log(D)> + B^2 <\log^2(D)> \tag{4.15}$$

where

$$<V> = \int_0^\infty V C(V) \, dV$$

and

$$<V^2> = \int_0^\infty V^2 C(V) \, dV \tag{4.16}$$

while

$$<\log(D)> = \frac{\int_0^\infty \log(D) C(V) \, dV}{\int_0^\infty C(V) \, dV}$$

and

$$<\log^2(D)> = \frac{\int_0^\infty \log^2(D) C(V) \, dV}{\int_0^\infty C(V) \, dV} \tag{4.17}$$

Though not yet theoretically proved, it is usually assumed that the differential area C(V)dV under a GPC curve is proportional to the differential mass of the polymers dW that are contained in the differential elution volume dV. Since $dW \propto f_w(M) dM$, we have

$$C(V) dV \propto f_w(M) dM \tag{4.18}$$

If C(V) is normalized, this gives

$$C(V) dV = f_w(M) dM \tag{4.19}$$

Combining Eqs. (4.3), (4.4) and (4.19), Eq. (4.17) can be rewritten

$$<\log(D)> = \frac{\int_0^\infty \log(D) G(D) D^{1/\alpha_D} dV}{\int_0^\infty G(D) D^{1/\alpha_D} dV}$$

and

$$<\log^2(D)> = \frac{\int_0^\infty \log^2(D)G(D)D^{1/\alpha_D}dV}{\int_0^\infty G(D)D^{1/\alpha_D}dV} \qquad (4.20)$$

After Eq. (4.19) is multiplied by M, both sides are integrated over the entire range of M to yield

$$M_w = \int_0^\infty MC(V)dV \qquad (4.21)$$

Elimination of M from the right-hand side using the relation $D = k_D M^{-\alpha_D}$ and Eq. (4.12) transforms Eq. (4.21) to

$$M_w = k_d^{1/\alpha_D} \int_0^\infty 10^{(A-V)/(\alpha_D B)} C(V) \, dV \qquad (4.22)$$

which is combined with Eq. (4.6) for M_w to give

$$\left[\int_0^\infty 10^{(A-V)/(\alpha_D B)} C(V) \, dV \right] \left[\int_0^\infty G(D) D^{1/\alpha_D} dD \right] = 1 \qquad (4.23)$$

This equation contains only one unknown parameter α_D. For a chosen α_D, we first calculate $<\log(D)>$ and $<\log^2(D)>$ using Eq (4.20), then calculate **A** and **B** by solving Eqs.(4.14) and (4.15) with $<V>$ and $<V^2>$ computed from the GPC diagram, and finally calculate the left side of Eq. (4.23). By iteration, we can find a value of α_D which may minimize the difference between the left and right sides of Eq. (4.23). For the α_D so obtained, we can calculate k_D from either Eq. (4.6) or (4.22) by using M_w determined directly from static LLS and C(V) from SEC or G(D) from dynamic LLS. With **A**, **B**, k_D and α_D, we are ready to calculate A and B. In this way, we can calibrate not only the M vs V relation, but also the M vs D by a single process on only one broadly distributed sample. This method has been tested and applied in the characterization of gelatin in water [50, 51].

5
Applications

When the relation between D and M is established, we can easily convert G(D) obtained by dynamic LLS into a differential molecular weight distribution, such as $f_w(M)$. We have successfully applied the above methods to various kinds of polymeric and colloidal systems, such as for Kevlar [15, 23], fluoropolymers (Tefzel & Teflon) [12, 30–35, 52], epoxy [53–55], polyethylene [56, 57], water-soluble polymers [18, 50–51, 58, 59], copolymers [60–62], thermoplastics [63–65] and colloids [66–72]. Three of those applications are illustrated below.

5.1
Segmented Copolymers

We consider a copolymer sample consisting of monomers A and B. The sample is generally polydisperse in both molecular weight and chain composition. We suppose that the copolymer species i is characterized by the molecular weight M_i and the composition w_{Ai} which is the weight fraction of A in the chain of that species. It is assumed that no composition heterogeneity exists in the chains of the same length. For homopolymers the refractive index increment (at infinite dilution) does not depend, in a good approximation, on the molecular weight of the chain, but is equal to that of the entire sample. For copolymers this is not the case, and, according to the theory of light scattering, Eq. (4.1) for G_i may be replaced by

$$G_i = (v_i/v)^2 M_i w_i / \sum_j M_j w_j \qquad (5.1)$$

where v_i is the refractive index increment due to the copolymer species i, v that of the entire sample, and w_i the weight fraction of the copolymer species i. When the molecular weight distribution may be treated as continuous. Equation (5.1) is generalized to

$$G(D)dD = (v(M)/v)^2 M f_w(M) M dM \qquad (5.2)$$

where, as before, $f_w(M)$ denotes the weight distribution of M, and $v(M)$ is the refractive index increment due to the chains of molecular weight M. Note that $v(M)$ depends on $w_A(M)$, which is the continuous version of w_{Ai}. We assume that this dependence is represented by

$$v(M) = v_A w_A(M) + v_B[1 - w_A(M)] \qquad (5.3)$$

where v_A and v_B are the refractive index increments of the homopolymers consisting respectively of A and B.

Corresponding to Eq. (4.3) for homopolymers, we introduce $f_{w,app}(M)$, an apparent weight distribution function of M, by

$$f_{w,app}(M) = (M_w/M)G(D)(dD/dM) \qquad (5.4)$$

With Eq. (4.4) (assumed to hold for copolymers too), this gives

$$f_{w,app}(M)/M_w = (1/k_D)^{2/\alpha_D} G(D) D^{1+(2/\alpha_D)} \qquad (5.5)$$

Therefore, $f_{w,app}(M)/M_w$ can be calculated from DLS determination of G(D) along with Eq. (4.4) when k_D and α_D are known separately. On the other hand, substituting Eq. (5.2) together with Eq. (5.3) into Eq. (5.4), we get

$$f_{w,app}(M)/M_w = v^{-2}[v_A w_A(M) + v_B(1 - w_A(M))]^2 f_w(M) \qquad (5.6)$$

Now, we choose two solvents 1 and 2 for a given copolymer. Since $f_{w,app}(M)$, v, v_A, and v_B vary with solvent, their values in solvent i are denoted by $f_{w,app}(M)^{(i)}$, $v^{(i)}$, $v_A^{(i)}$, and $v_B^{(i)}$. Then it follows from Eq. (5.6) that

$$\frac{f_{w,app}^{(1)}(M)}{f_{w,app}^{(2)}(M)} = \left\{ \frac{v^{(2)}}{v^{(1)}} \frac{w_A(M)v_A^{(1)} + [1 - w_A(M)]v_B^{(1)}}{w_A(M)v_A^{(2)} + [1 - w_A(M)]v_B^{(2)}} \right\}^2 \qquad (5.7)$$

The ratio on the left-hand side can be obtained as a function of M since $f_{w,app}(M)^{(i)}/M_w$ can be determined, as described above, from experimental information. Thus, Eq. (5.7) allows determination of $w_A(M)$, the chain composition distribution when all other parameters on its right-and side are measured by differential refractometry. Once $w_A(M)$ is known, we are ready to compute $v(M)$ from Eq. (5.3), $f_w(M)$ from Eq. (5.6), and finally M_w [59].

Figure 14 shows PLS determined – G(D)s by Eq. (5.3) for low-mass (○) and high-mass (□) segmented copolymer poly(ethylene terephthalate-co-caprolactone)s (PET-PCL) containing 13% PET in tetrahydrofuran (THF) at 25 °C. Repeating the measurements on these samples in another solvent chloroform should lead to a new set of G(D)s, which allows Eq. (5.7) to be used to calculate $w_{PET}(M)$.

Figure 15 shows the results from such calculations for low mass (○) and high mass (□)13% PET-PCL samples. We see that the PET content increases with increasing M for M<~4×10⁴ and approaches a constant value of ~14% in the high molar mass range. For the 58% PET-PCL sample, the composition distribution is nearly constant. The composition of the high-molar mass 13% PET-PCL sample overlaps with that of the low mass 13% PET-PCL sample in the same molec-

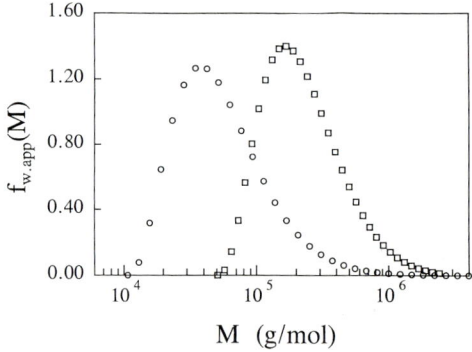

Fig. 14. Apparent weight distributions calculated from the translational diffusion coefficient distributions corresponding to low-mass (○) and high-mass (□) copolymer segmented poly(ethylene terephthalate-co-caprolactone) (PET-PCL) containing 13% PET in tetrahydrofuran (THF) at 25 °C

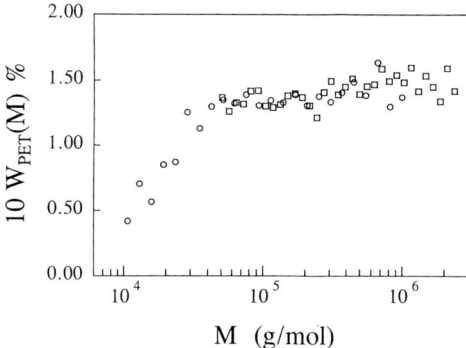

Fig. 15. Estimate of the chain composition distributions for low mass (○) and high mass (□) 13% PET-PCL samples

ular weight range. This indirectly indicates that the estimation of the composition distribution is reasonable. The lower content of PET in the low molar mass range can be attributed to the two-step synthesis [59–62].

5.2
A Polymer Mixture Containing Individual Chains and Clusters

If a mixture is made of individual polymer chains and polymer clusters, the measurement of static LLS will lead to an apparent weight-average molar mass $M_{w,app}$ which is expressed by

$$M_{w,app} = M_{w,L}x_L + M_{w,H}x_H \tag{5.8}$$

where the subscripts "L" and "H" denote low molar mass linear polymer chains and high molar mass polymer clusters, respectively, and x_L and x_H are their weight fractions with $x_L+x_H=1$. If the linear chains and clusters are significantly different in the hydrodynamic size, dynamic LLS will detect two distinct peaks in the measured $G(D)$, with one peak corresponding to the linear chains and the other to clusters.

Figure 16 shows $G(D)$ of a simulated polymer mixture at two scattering angles ("○", 14° and "□", 17°). The mixture consists of two polystyrene standards having distinctly different weight average molar masses (3.0×10^5 and 5.9×10^6 g/mol) and a high mass polystyrene which is used to simulate the polymer cluster [66]. The area ratio A_r of the two peaks is expressed by

$$A_r = \frac{A_L}{A_H} = \frac{\int_0^\gamma G_L(D)\, dD}{\int_\gamma^\infty G_H(D)\, dD} = \frac{M_{w,L}x_L}{M_{w,H}x_H} \tag{5.9}$$

with γ the cutoff translational diffusion coefficient between $G_L(D)$ and $G_H(D)$. In practice, the values of A_r at finite scattering angles must be extrapolated to

$q \to 0$. Figure 17 shows this extrapolation of the A_r for the two peaks in Fig. 16. With Eqs. (5.8) and (5.9), $M_{w,app}$ from static LLS and A_r from dynamic LLS allow $M_{w,L}x_L$ and $M_{w,H}x_H$ to be computed. In principle, by knowing any one of the four parameters ($M_{w,L}$, $M_{w,H}$, x_L and x_H), we should be able to determine the remaining three parameters. This method has been thoroughly tested with mixtures of polystyrene standards [64]. As for a particular polymer mixture, we should find a way allowing determination of one of the four parameters. For example, in the study of polymer association, we can determine the $M_{w,L}$ of starting individual polymer chains, and in the study of the gelation process, we can use a filtration method to remove large microgels, so that the weight fractions of x_L and x_H can be subsequently determined. This method has been used to characterize novel thermoplastic polymers with phenolphthalein in their backbone chains [61–63].

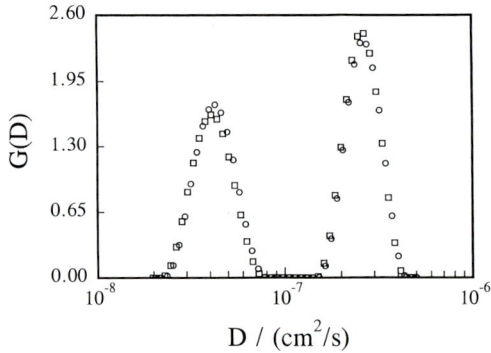

Fig. 16. Translational diffusion coefficient distributions G(D) of a simulated polymer mixture at two scattering angles ("□", 17° and "○", 14°). The mixture contains two polystyrene standards of distinctly different weight average molar masses (3.0×10^5 and 5.9×10^6 g/mol) and a high mass polystyrene

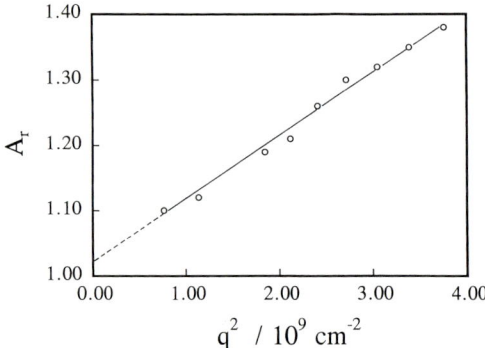

Fig. 17. q^2-dependence of the area ratio ($A_r = A_L/A_H$) for the two peaks of the translational diffusion coefficient distribution shown in Fig. 16

5.3
Polymer Colloids

Combination of static and dynamic laser light scattering is also useful to determine not only the size distribution but also the particle structure of polymer colloids such as the adsorbed surfactant layer thickness [73] and the formation of nanoparticles [74, 75]. A recently developed method of determining the density of polymer particles is outlined below to illustrate the usefulness of laser light scattering as a powerful analytical tool for investigating more sophisticated colloidal problems [76–78].

For a colloidal particle of uniform density its molar mass M is proportional to the cube of its radius R, i.e.

$$M = (4/3)\pi R^3 \varrho N_A \qquad (5.10)$$

where ϱ is the particle density and N_A the Avogadro constant. The diffusion coefficient D of the particle (at infinite dilution) is related to the Stokes radius R_h by

$$D = (k_B T / 6\pi\eta)(1 / R_h) \qquad (5.11)$$

where k_B is the Boltzmann constant, T the absolute temperature, and η the solvent viscosity. We assume that R_h is larger than R by the thickness b of the solvated layer, i.e.

$$R_h = R + b \qquad (5.12)$$

Substituting this into Eq. (5.11) and using Eq. (5.10) for R, we obtain

$$D = F(k_B T/6\pi\eta)(4\pi\varrho N_A)^{1/3} M^{-1/3} \qquad (5.13)$$

with

$$F = 1/[1+b(4\pi\varrho N_A/M)^{1/3}] \qquad (5.14)$$

Comparing Eq. (5.13) with the relation $D = k_D M^{-\alpha_D}$ and considering $b \ll R$, we find approximately

$$\alpha = 1/3 \qquad (5.15)$$

$$k_D = F(k_B T/6\pi\eta)(4\pi\varrho N_A)^{1/3} \qquad (5.16)$$

Thus, with M in Eq. (5.14) replaced by M_w, it follows from Eq. (4.6) that

$$M_w = \frac{1}{[1+b(4\pi\rho N_A / M_w)^{1/3}]^3} \left(\frac{4\pi N_A}{3}\right)\left(\frac{k_B T}{6\pi\eta}\right)^3 \bigg/ \int_0^\infty G(D) D^3 dD \qquad (5.17)$$

This equation contains two unknown parameters (b and ϱ), and if we know one of them, the other can be calculated from M_w and G(D). With this idea, it was found that the average density of the polystyrene microspheres made of a few uncrosslinked chains is slightly lower than that of bulk polystyrene or conventional polystyrene latex.

6
Conclusions

This review has shown that static and dynamic laser light scattering (LLS) combined provide a very powerful method for polymer characterization. LLS has advantages over other polymer characterization methods, which include ultracentrifugation and chromatography, in such features as speed, non-perturbation and extreme dissolution conditions (high temperature or strong acid). The most important advantage is that the calibration is independent of the particular LLS instrument used. However, the LLS method for the determination of mass distributions described in this paper is disadvantageous in that its resolution is not as high as the fractionation methods, especially for samples whose mass distributions have closely packed peaks. The LLS method should play a definite role in circumstances where polymers intractable by conventional characterization methods have to be treated. In principle, dynamic LLS can be used together with other polymer characterization methods which take advantage of the dependence of the hydrodynamic volume on molecular weight.

Acknowledgments. My special thanks to Professor Benjamin Chu who led me to the laser light scattering field fourteen years ago. I am also indebted to my former colleagues at Stony Brook (USA) and in BASF (Germany) for their cooperation, to my collaborators in China for their help, and to my postgraduate students for their dedication to our research projects in the past years. The financial support of the Research Grants Council of Hong Kong Government from 1993 to 1996 is gratefully acknowledged. I would also like to thank Professor Hiroshi Fujita for his constant guidance and encouragement in the process of writing this review.

References and Notes

1. Flory PJ, Krigbaum WR (1950) J Chem Phys 18:1086
2. Grimley TB (1952) Proc Roy Soc (London) A212:339
3. Chien JY, Shih LH, Yu SC (1958) J Polym Sci 26(119):117
4. Huglin MB (1972) Light scattering from polymer solution. Academic, London
5. Chu B, Fytas G (1982) Macromolecules 15:562
6. Ford NC (1983) In: Dahneke B (ed) Measurement of suspended particles by quasi-elastic light scattering. Wiley, New York
7. Pecora R (ed) (1985) Dynamic light scattering. Plenum, New York
8. Brown W (ed) (1996) Light scattering: principles and development. Clarendon, Oxford
9. Chu B (1991) Laser light scattering. Academic. New York, 2nd edn
10. Berne BJ, Pecora R (1976) Dynamic light scattering. Plenum, New York
11. Zimm BH (1948) J Chem Phys 16:1099
12. Chu B, Wu C (1987) Macromolecules 20:2642
13. Russo PS (1993) In: Brown W (ed) Dynamic light scattering, Chap 12. Oxford University Press, New York

14. Chu B, Ying Q, Wu C, Ford J (1984) Polymer Communications 25:211
15. Chu B, Ying Q, Wu C (1985) Polymer 26:1408
16. Ying Q, Chu B (1984) Die Makromolekulare Chemie, Rapid Communication 5:785
17. Stockmayer WH, Schmidt M (1984) Macromolecules 17:509
18. Wu C, Zhou SQ, Wang W (1995) Biopolymer 35:385
19. Koppel DE (1972) J Chem Phys 57:4814
20. Livesey AK, Licinio P, Delaye M (1986) J Chem Phys 84:5102
21. McWhirter JG, Pike ER (1978) J Phys A 11:1729
22. Chu B, Ford JR, Dhadwal HS (1983) Methods Enzymol 117:256
23. Chu B, Wu C, Ford JR (1985) J Colloid Interface Sci 105:473
24. Provencher SW (1976) J Chem Phys 64:2772
25. Chu B, Wu C, Ford JR (1985) J Colloid Interface Sci 105:473
26. Bushuk W, Benoit H (1958) Can J Chem 36:1616
27. Chu B, Ying Q, Lee DC, Wu DQ (1985) Macromolecules 18:1962
28. Stockmayer WH, Moore LD, Fixman M, Epstein BN (1955) J Polym Sci 16:517
29. Wu C, Woo KF, Luo XL, Ma DZ (1994) Macromolecules 27:6055
30. Chu B, Wu C (1987) Macromolecules 20:93
31. Wu C, Buck W, Chu B (1987) Macromolecules 20:98
32. Chu B, Wu C, Zuo J (1987) Macromolecules 20:700
33. Chu B, Wu C, Buck W (1988) Macromolecules 21:397
34. Chu B, Wu C, Buck W (1989) Macromolecules 22:831
35. Wu C (1992) Makromol Chem, Makromol Symp 61:377
36. Obasa M, Nakamura H, Takasaka M, Keto T, Nagasawa M (1993) Polym J 25:301
37. Application Notes LS7 Chromatix, Mountain View, California, USA
38. Baltog I, Ghita C, Ghita L (1970) Eur Polym J 6:1299
39. Application Notes Optilab 903, Wyatt Technology, Santa Barbara, California, USA
40. Wu C, Xia KQ (1994) Review of Scientific Instruments 65:587
41. de Gennes PG (1979) Scaling concepts in polymer physics. Cornell University Press, Ithaca, NY
42. Raczek J (1983) Eur Polym J 19:607
43. Nordmeier E, Lechner MD (1989) Polym J 21:623
44. Appelt B, Meyerhoff G (1980) Macromolecules 13:657
45. Wu C, Zhang YB, Yan XH, Cheng RS (1995) Acta Polymerica Sinica 3:349
46. Flory PJ (1953) Principles of polymer chemistry. Cornell University Press, Ithaca, NY
47. Chu B, Onclin M, Ford JR (1984) J Phys Chem 88:6566
48. Pope J, Chu B (1984) Macromolecules 17:2633
49. Cervenka A (1973) Makromol Chem 170:239
50. Wu C (1993) Macromolecules 26:5423
51. Wu C (1994) J Polym Sci Polym Phys Ed 32:803
52. Wu C, Chu B (1986) Macromolecules 19:1285
53. Wu C, Zuo J, Chu B (1989) Macromolecules 22:633
54. Wu C, Zuo J, Chu B (1989) Macromolecules 22:838
55. Wu C, Chu B, Stell G (1992) Makromol Chem, Makromol Symp 45:75
56. Wu C, Lilge D (1993) J Appl Polym Sci 50:1753
57. Wu C (1993) J Appl Polym Sci 54:969
58. Zhou SQ, Fan SF, Au-yeung SCF, Wu C (1995) Polymer 36:1341
59. Wu C, Wu PQ, Ma XQ (1995) J Polym Sci Polym Phys Ed, in press
60. Wu C, Ma D, Luo X, Chan KK, Woo KF (1994) J Appl Polym Sci 53:1323
61. Wu C, Woo KF, Luo X, Ma D (1994) Macromolecules 27:6055
62. Woo KF, Wu C (1995) J Appl Polym Sci 57:1285
63. Bo S, Siddiq M, Wu C (1996) Macromolecules 29:2989
64. Wu C, Siddiq M, Bo B (1995) Macromolecules 29:3157
65. Siddiq M, Li BY, Wu C (1996) J Appl Polym Sci 60: 1995
66. Wu C, Siddiq M, Woo KF (1995) Macromolecules 28:4914
67. Wu C (1994) Macromolecules 27:298

68. Wu C (1994) Macromolecules 27:7099
69. Wu C, Qian RY, Napper DH (1995) Macromolecules 28:1592
70. Zhou SQ, Wu C (1995) Macromolecules 28:5225
71. Wu C (1994) Chinese J Polym Sci 12:323
72. Wu C, Chan KK (1995) J Polym Sci Polym Phys Ed 33:919
73. Wu C (1994) Macromolecules 27:298, 7099
74. Li M, Jiang M, Wu C (1997) Macromolecules 30:2201
75. Wu C, Chen M, Akashi M (1997) Macromolecules 30:2187
76. Wu C, Chan KK (1994) J Polym Sci, Polym Phys Ed 33:919
77. Wu C, Qian RY, Napper DH (1994) Macromolecules 28:1592
78. Gao J, Zhou SQ, Wu C (1996) Polym Engineering Sci 36:2968

Editor: Prof. Hiroshi Fujita
Received: September 1997

Mean-Field Kinetic Modeling of Polymerization: The Smoluchowski Coagulation Equation*

Henryk Galina[1] and Jaromir B. Lechowicz

[1] Department of Industrial and Materials Chemistry, Faculty of Chemistry, Rzeszów University of Technology, 35-959 Rzeszów, Poland. E-mail: chemia@prz.rzeszow.pl

The Smoluchowski coagulation equation derived back in 1916 is usually linked with the diffusivity and size of aggregating particles. It can also be used as a versatile tool for mean-field kinetic analyses. In this paper it is shown to be an efficient tool for studying the relationships between the reactivity of functional groups in monomers and the size distribution of polymer species in step and chain growth polymerizations. The Smoluchowski coagulation equation and its modifications are applied as models of kinetically controlled growth reactions (with irreversible elementary reaction steps). Advantages and limitations of this method of modeling polymerization processes are discussed.

Keywords: Size distribution, cross-linking polymerization, aggregation, classical modeling, network formation, gelation

1	Introduction	136
2	The Smoluchowski Coagulation Equation	137
3	Random Step Growth Homopolymerization of a Bifunctional Monomer	139
4	Homopolymerization of a Bifunctional Monomer Reacting with Substitution Effect	143
5	Multicomponent System of Bifunctional Monomers	145
6	Random Homopolymerization of a Monomer with More than Two Functional Groups	151
7	The Equivalence Between Kinetics and Simple Statistical Models	155
8	Gelling Systems with Substitution Effect	156
9	Post-Gelation Relationships	162
10	Attempts at Taking Into Account Cyclization Reactions	165
11	Other Remarks on the Smoluchowski Coagulation Equation	168
12	References	170

---

* Affectionately dedicated to Professor Manfred Gordon on his 80th birthday

List of Symbols and Abbreviations

A, B	types of functional groups
c_k	concentration of k-mers
d	space dimensionality
ΔE	activation energy
f	functionality of a unit
$H, G, E, \tilde{A}, \tilde{B}$	generating (counting) functions
$K_{i,j}$	coagulation kernel, the rate constant for the reaction between i- and j-mer
$k_{i,j}$	rate constant for the reaction between i- and j-mer
M_k	k-th moment of size distribution ($k=0,1,2,\ldots$)
N	total number of units in the system
N_A	the Avogadro constant
n_k	number (mol) fraction of k-mers
p	conversion degree
P_1	aggregate of size i, polymer molecule of polymerization degree i, i-mer
P_n	number average degree of polymerization
P_w	weight average degree of polymerization
$RA_2, R'B_2$, ARB, RA_∞	types of monomers (A and B denote functional groups)
s_i	component of the product kernel
t	time
t_c	reaction time at the gel point
w_k	number of ways of assembling a k-mer
x, y, z	dummy variables in generating functions
α, β, A, B, C	numerical constants
κ, k_A, k_B	ratios of rate constants, relative rate constants
λ	cyclization parameter, degree of homogeneity,
ψ	contribution to activation energy
σ	symmetry factor
τ	units proportional to time, rescaled time
ξ, ζ	parameters in generating functions beyond the gel point
$\omega, \alpha, \theta, \tau, z$	exponents

1
Introduction

The modeling of a polymerization process is usually understood as formulation of a set of mathematical equations or computer code which are able to produce information on the composition of a reacting mixture. The input parameters are reaction paths and reactivities of functional groups (or sites) at monomeric substrates. The information to be modeled may be the averages of molecular weight, mean square radius of gyration, particle scattering factor, moduli of elasticity, etc. Certain features of polymerizations can also be predicted by the models.

Among them is the gel point conversion, if multifunctional units are present, as well as accompanying divergence of viscosity, onset of equilibrium elasticity modulus, etc. By comparing the results of modeling with experiment, one can verify to what extent the chemistry is affected by physical interactions which are practically always active in polymerizations.

Note that all the output parameters mentioned are directly related to system connectivity. The meaning of the word *chemistry* is in fact reduced to connectivity build-up since each chemical reaction between two functional groups leads to the formation of a link.

This paper deals with one of the mean-field methods of modeling the connectivity build-up that can be applied to polymerization processes. As in the other mean-field methods of modeling, certain physical features such as concentration fluctuations or fluctuation coupled diffusion control of reaction steps, etc., are neglected.

The paper concentrates on the use of the Smoluchowski coagulation equation which is presented in Sect. 2. This equation is usually linked to the effects of diffusion and size of colloid particles in aggregation processes. Here, however, we refer to it as to probably the first known kinetic model applicable for the analysis of parallel-consecutive reactions. In Sects. 3–5 the Smoluchowski coagulation equation is shown to be a useful tool for kinetic modeling the step growth polymerization of one or copolymerization of two bifunctional monomers. Section 6 describes the use of the Smoluchowski equation to model the polymerization of multifunctional monomers. The equivalence of the model and those based on assemblage of structures from fragments using statistical arguments is discussed in Sect. 7. An example of modeling a kinetically-controlled polymerization process which yields results differing from those obtained by using statistical approaches is presented in Sect. 8. Post-gelation analysis which is available for the simplified systems only is described in Sect. 9. In Sect. 10 attempts at taking into consideration intramolecular reactions are reviewed. Finally, in Sect. 11, some general features of the Smoluchowski coagulation equation, mostly those loosely related to polymer science, are presented.

2
The Smoluchowski Coagulation Equation

Consider an irreversible second order aggregation process where an aggregate of size (or mass) i and an aggregate of size (mass) j combine to form an $i+j$ aggregate:

$$P_i + P_j \xrightarrow{K_{i,j}} P_{i+j} \quad (i,j=1,2,...) \tag{1}$$

The size of aggregates is an integer quantity. It is therefore convenient to consider it as a multiplicity of the size of the unit aggregate P_1, or, if applied to polymers, as the polymerization degree. Adopting polymer terminology, we shall refer to an aggregate of size i as an i-mer.

To characterize the process defined by Eq. (1), Smoluchowski applied the empirical [1] Guldberg-Waage [2] *law of mass action*. In his papers [3,4] published in 1916 and 1917 he derived and examined the set of differential kinetic equations equivalent to

$$\frac{dc_1}{dt} = -c_1 \sum_{i=1}^{\infty} K_{1,i} c_i$$

$$\frac{dc_2}{dt} = \tfrac{1}{2} K_{1,1} c_1^2 - c_2 \sum_{i=1}^{\infty} K_{2,i} c_i \qquad (2)$$

$$\frac{dc_3}{dt} = K_{1,2} c_1 c_2 - c_3 \sum_{i=1}^{\infty} K_{3,i} c_i$$

Nowadays, the set at Eq. (2) is known better in its compact form

$$\frac{dc_k}{dt} = \tfrac{1}{2} \sum_{i=1}^{k-1} K_{i,k-i} c_i c_{k-i} - c_k \sum_{i=1}^{\infty} K_{k,i} c_i \qquad (3)$$

which is called the *Smoluchowski coagulation equation*.

Solutions of the Smoluchowski equation are of interest in all branches of physics where aggregation processes take place. There is a vast literature on its application including several reviews [5-7]. Here, we concentrate on those applications which can be used to model the polymerization processes.

The $K_{i,j}$ in Eqs. (1)-(3) is the rate constant of aggregation. The set of rate constants for all i and j, often called the *coagulation kernel*, is an infinite symmetric matrix with nonnegative elements. In order to keep the form of $K_{i,j}$ as simple as possible we shall assume that

- the system is closed, i.e., the total number of units, N, is constant,
- the system is uniform with no concentration fluctuations affecting the reaction,
- $K_{i,j}$ depends only on i and j.

The concentration of a k-mer, c_k, can be expressed in arbitrary units. It is convenient, however, to express its concentration as

$$c_k = \frac{\text{number of } k\text{-mers}}{N} \qquad (4)$$

The set of c_k's constitutes the distribution of aggregate sizes. The Smoluchowski equation (Eq. 3) defines its change with time.

In any kinetic analysis, the time, t, comes naturally as an independent variable, but in Eq. (3) it is, in fact, a variable merely proportional to the real time. It can be regarded as a rescaled time with the scaling factors depending on the actual units of both coagulation kernel and concentration as well as on the type of polymerization reaction considered. Suppose we changed the definition at Eq. (4) of concentration to become simply the number of moles of k-mers. Then, the new time units would be related to the old ones with the scaling factor N/N_A (N_A is the Avogadro constant).

The form of the Smoluchowski coagulation equation at Eq. (3) is significant. The first term on the right hand side describes the rate at which k-mers are formed from smaller components of matching sizes. The factor 1/2 takes care of the symmetry. It prevents the terms $K_{j,i}c_jc_i$ and $K_{i,j}c_ic_j$ both being counted. For $i=j$, the 1/2 reflects the fact that formation rate of $2i$-mer is only a half of the rate at which i-mers disappear from the system [8]. The second negative term of Eq. (3) is the rate at which k-mers disappear from the system in *reactions of growth*. Since the aggregation is irreversible, there is no term for the decomposition of k-mers into smaller components. Unless otherwise stated, we will consider formation of acyclic (tree-like) aggregates only. In polymer parlance, intramolecular reactions will be disregarded.

The definition at Eq. (4) implies that the zeroth moment of the k-mer size distribution, M_0, is the reciprocal of the number average degree of polymerization in the system

$$M_0 = \sum_{i=1}^{\infty} c_i = \frac{1}{P_n} \qquad (5)$$

and the first moment provides the normalization of the distribution

$$M_1 = \sum_{i=1}^{\infty} ic_i = 1 \qquad (6)$$

In the following sections it will be shown how the Smoluchowski equation, or rather its modifications, can be used to predict these two and higher moments of size distributions for various polymer systems.

3
Random Step Growth Homopolymerization of a Bifunctional Monomer

This classical polymerization model [9] is sometimes referred to as an RA_2 or ARB system, with A and B denoting functional groups. The groups may react either with themselves (RA_2 system) or A groups may react with B groups only (ARB system). The 'linear' polymerization described in this section is in fact a special case of models 1 or 3 of Table 4 in Sect. 6, but it is dealt with separately because of its special importance to polymer science.

The functional groups in reacting species (i-mers) are equally reactive irrespectively of i. The reaction between an i-mer and j-mer can be written as

$$P_i + P_j \xrightarrow{\sigma k_{i,j}} P_{i+j} \qquad (i,j=1,2,..) \qquad (7)$$

Since $k_{i,j}$ does not depend on i or j, it can be absorbed, together with the symmetry factor σ (equal to 4 for RA_2 or 2 for ARB), into time units and the Smoluchowski coagulation equation would assume its simplest form with the kernel equal to one:

$$\frac{dc_k}{dt} = \frac{1}{2}\sum_{i=1}^{k-1} c_i c_{k-i} - c_k \sum_{i=1}^{\infty} c_i \qquad (8)$$

The solution of Eq. (8) was presented by Smoluchowski himself [4]. The time dependence of the zeroth moment of the distribution is particularly easy to calculate from Eq. (8). Simply sum up the terms in Eq. (8) for all k's to obtain

$$\frac{dM_0}{dt} = \frac{1}{2}M_0^2 - M_0^2 = -\frac{1}{2}M_0^2 \qquad (9)$$

since

$$\sum_{k=1}^{\infty}\sum_{i=1}^{k-1} c_i c_{k-i} = \left(\sum_{i=1}^{\infty} c_i\right)^2 = M_0^2 \qquad (10)$$

The solution of Eq. (9) is

$$M_0(t) = \frac{2}{t+2} \qquad (11)$$

A convenient method of extracting higher moments of the distribution directly from Eq. (8) (to be fully exploited in following sections) is to define the polynomial generating function [10] containing concentrations of species, $c_i = c_i(t)$, and the dummy variable x with no physical significance

$$H(t,x) = \sum_{i=1}^{\infty} c_i x^i \qquad (12)$$

Now, multiplication of both sides of Eq. (8) by x^k followed by summing up for all k's yields

$$\frac{\partial H}{\partial t} = \frac{1}{2}H^2 - HM_0 \qquad (13)$$

The solution of Eq. (13) is

$$H = \frac{4x}{(t+2)\left[t(1-x)+2\right]} \qquad (14)$$

Note, however, that since the successive moments of distribution are simply given by

$$\left.\frac{\partial H}{\partial x}\right|_{x=1} = M_1; \quad \left.\frac{\partial^2 H}{\partial x^2}\right|_{x=1} = M_2 - M_1; \quad \left.\frac{\partial^3 H}{\partial x^3}\right|_{x=1} = M_3 - 3M_2 + 2M_1 \quad (15)$$

etc., we can convert Eq. (13) without even solving it into a set of ordinary differential equations for the moments. Thus, by differentiating Eq. (13) with respect to x followed by substituting $x=1$ we obtain

$$\frac{\partial^2 H}{\partial t \partial x} = H \frac{\partial H}{\partial x} - M_0 \frac{\partial H}{\partial x} \Rightarrow \frac{dM_1}{dt} = 0 \Rightarrow M_1 = 1 \quad (16)$$

since $H(t,1)=M_0$,

$$\frac{\partial^3 H}{\partial t \partial x^2} = \left(\frac{\partial H}{\partial x}\right)^2 + H \frac{\partial^2 H}{\partial x^2} - M_0 \frac{\partial^2 H}{\partial x^2} \Rightarrow \frac{d(M_2 - M_1)}{dt} = M_1^2 \Rightarrow M_2(t) = t+1 \quad (17)$$

$$\frac{\partial^4 H}{\partial t \partial x^3} = 3 \frac{\partial H}{\partial x} \frac{\partial^2 H}{\partial x^2} + H \frac{\partial^3 H}{\partial x^3} - M_0 \frac{\partial^3 H}{\partial x^3} \Rightarrow \frac{d(M_3 - 3M_2 + 2M_1)}{dt} =$$
$$3M_1(M_2 - M_1) \Rightarrow M_3(t) = \tfrac{1}{2}(3t^2 + 6t + 2) \quad (18)$$

etc.

From the moments we can calculate the number, weight, and z-average degrees of polymerization which are defined thus:

$$P_n = M_1/M_0; \quad P_w = M_2/M_1; \quad P_z = M_3/M_2 \quad (19)$$

respectively. The first averages of polymerization degree calculated from Eq. (19) are listed in Table 1. They are shown as functions of (rescaled) reaction time as well as functions of conversion degree, p. The latter is, as usual, defined as the fraction of all functional groups in the system that has reacted to bonds. The relation between conversion and time is derived by noting that two functional groups have to react to form a bond and that each k-mer has exactly $k-1$ bonds. Hence

Table 1. Number, weight, and z-average degrees of polymerization for the random homopolymerization of bifunctional monomer calculated using the Smoluchowski equation (Eq. 8) and expressed in terms of time (t) and conversion degree (p)

P_n	$\frac{1}{2}t + 1$	$\frac{1}{1-p}$
P_w	$t + 1$	$\frac{1+p}{1-p}$
P_z	$\frac{1}{2}(3t^2 + 6t + 2)$	$\frac{1 + 4p + p^2}{1 - p^2}$

$$p = \frac{2N \sum_{k=1}^{\infty}(k-1)c_k}{2N} = M_1 - M_0 \qquad (20)$$

or

$$p = \frac{t}{t+2} \qquad (21)$$

By using Eq. (21) we find out that the moments of the distribution obtained by using the Smoluchowski equation are exactly the same as those derived for the same random homopolymerization using statistical arguments [9].

The explicit distribution of molecular sizes can now be obtained by solving recursively the Smoluchowski equation at Eq. (8). For c_1 we get

$$\frac{dc_1}{dt} = -c_1 M_0 \qquad (22)$$

which instantly gives

$$c_1 = \frac{4}{(t+2)^2} \qquad (23)$$

Similarly, by solving

$$\frac{dc_2}{dt} = \frac{1}{2}c_1^2 - c_2 M_0 \qquad (24)$$

we obtain

$$c_2 = \frac{4t}{(t+2)^3} \qquad (25)$$

etc. It is not difficult to show that carrying out with solving consecutive equations we would obtain the entire distribution [11]

$$c_k = \frac{4t^{k-1}}{(t+2)^{k+1}} \qquad (26)$$

(Obviously, the same distribution is obtained by expanding Eq. (14) into a power series around $x=1$.) This is the Flory most probable distribution. One can easily verify it by replacing c_k with number (mol) fraction of k-mers, n_k, by using the relation

$$c_k = n_k M_0 \qquad (27)$$

Then, by replacing time with conversion using Eq. (21) we arrive at the familiar [9]

$$n_k = (1-p)p^{k-1} \qquad (28)$$

Thus, for the random homopolymerization of a 2-functional monomer, the kinetic analysis leads to the same molecular size distribution as that derived by using statistical arguments. Conditions for such an equivalence will be discussed in the following sections

4
Homopolymerization of a Bifunctional Monomer Reacting with Substitution Effect

Many monomers react with the so called substitution effect. To deal with it quantitatively we shall adopt the approach of Gordon and Scantlebury [12]. Suppose the monomer molecule contains f similar functional groups and every group has, or has not, the same reactivity. The *zeroth shell substitution effect* is simply no effect at all. Each functional group reacts independently of whether or not other groups have reacted. If, however, the reactivity of remaining $f-1$ groups changes as soon as the first has reacted, we have the *first shell substitution effect* (FSSE). Clearly, with FSSE the reactivity of remaining $f-2, f-3,...$ unreacted groups may still change as the successive groups react. In other words, with FSSE the unit's substitution degree determines the reactivity of its unreacted groups. An example of the FSSE that is particularly appealing to chemists seems to be the way the primary amino group reacts. Initially, both amino protons have an equal chance of entering a reaction, but after one has reacted the reactivity of the other proton usually changes.

The *second shell substitution effect* (SSSE) is operative when the reactivity of a group depends not only on the substitution degree of the unit it belongs to, but on the substitution degree of all units that are first neighbors to that unit. Unless stated otherwise, we shall assume for simplicity that all the unreacted groups of a unit have an equal chance of reacting.

When an RA_2 monomer reacts with the substitution effect then the application of kinetics analysis will lead to a different molecular distribution to that resulting from a simple statistical analysis. This was convincingly demonstrated by Kuchanov [6]. For the kinetics analysis one has to distinguish three types of reaction:

$$P_1 + P_1 \xrightarrow{4k_{11}} P_2$$
$$P_1 + P_i \xrightarrow{4k_{12}} P_{i+1}$$
$$P_i + P_j \xrightarrow{4k_{22}} P_{i+j}$$
$$i, j = 2, 3, \ldots$$

(29)

The analysis becomes simpler if we assume, following Gordon and Scantlebury [13], that the contributions from reacting units to the activation energy of reaction between their reactive groups are additive. Let ΔE be the activation energy of the reaction between groups on unsubstituted units (monomers). When a monomer molecule reacts with a unit having one reactive group already reacted, let the activation energy change by the value of ψ. Clearly, when each of the reacting units has one of its groups already reacted then the activation energy will be $\Delta E + 2\psi$. With this assumption we may absorb the constant $4k_{22}$ into the time units and deal only with the relative rate constant defined thus:

$$\kappa = k_{12}/k_{22} = \exp\left(-\psi/RT\right)$$

(30)

so that

$$k_{11}/k_{22} = \kappa^2$$

(31)

The Smoluchowski equation becomes now slightly more complicated. It can be written in the form

$$\frac{dc_1}{dt} = -\kappa^2 c_1^2 - \kappa c_1 \sum_{i=2}^{\infty} c_i$$

$$\frac{dc_2}{dt} = \frac{1}{2}\kappa^2 c_1^2 - \kappa c_1 c_2 - c_2 \sum_{i=2}^{\infty} c_i$$

(32)

$$\frac{dc_k}{dt} = \frac{1}{2}\sum_{i=2}^{k-2} c_i c_{k-i} - c_k \sum_{i=2}^{\infty} c_i + \kappa c_1(c_{k-1} - c_k)$$

When multiplied by x^i and summed up over all k it becomes

$$\frac{\partial H}{\partial t} = \frac{1}{2}\tilde{H}^2 - \tilde{H}\tilde{H}_0$$

(33)

with

$$\tilde{H}(t,x) = \kappa c_1 x + \sum_{i=2}^{\infty} c_i x^i \qquad (34)$$

where

$$\tilde{H}_0(t) = \tilde{H}(t,1) \qquad (35)$$

Note that the generating function in Eq. (33) without the tilde is the same as that given by Eq. (12) when $x=1$. The complete equation at Eq, (33) also becomes Eq. (13) when $x=1$, but it is difficult to solve analytically for any x. The moments of the distribution, however, can be calculated numerically as functions of time since

$$\tilde{H} = H + (\kappa - 1) c_1 x \qquad (36)$$

By using the first line of Eq. (32) written in the form

$$\frac{dc_1}{dt} = -\kappa c_1 \tilde{H}_0 \qquad (37)$$

one may extract the moments by successive differentiation of Eq. (33) with respect to x followed by putting $x=1$.

Comparison of the weight average degrees of polymerization (equal to second moments) calculated by using Eq. (33) and a simple statistical model (cascade theory [14, 15]) is presented in Fig. 1. The nonequivalence of the two methods can clearly be seen. The number fractions of the monomer, once-, and doubly-reacted units were calculated from Eq. (33). The fractions were treated as probabilities for the units of being in their respective reaction states. The „statistical" weight average degree of polymerization was then calculated by using the probabilities following the cascade theory standard procedures [14, 15].

5
Multicomponent System of Bifunctional Monomers

An example of multicomponent system that can be dealt with by using kinetics equations of the Smoluchowski type is provided by a step growth alternating copolymerization of two bifunctional monomers [16]. This system requires little more laborious, but quite straightforward algebra.

Let the reaction involve monomers RA_2 and $R'B_2$. Functional groups A react with functional groups B. One can distinguish as many as nine elementary reaction steps. This is because each molecule contains either an even or odd number of units. Those with odd numbers of units may be terminated with either two

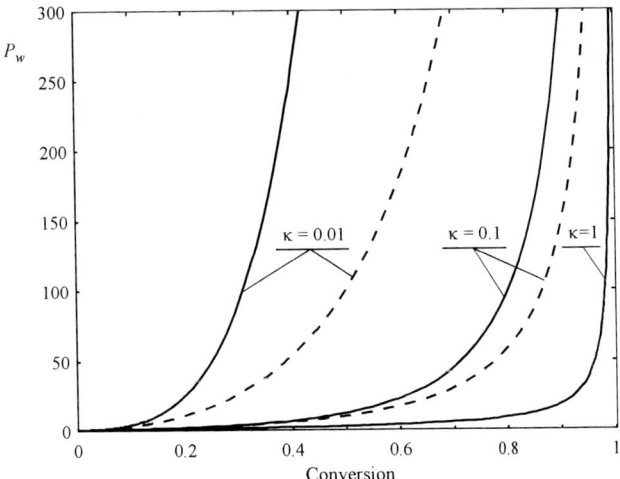

Fig. 1. The weight average degree of polymerization, P_w, in a linear polymerization with monomer reacting with a substitution effect. *Solid line* – calculated according to Eq. (33), *broken line* – calculated by using cascade theory for the specified relative rate constant k

groups A or two groups B and hence they may react differently. The monomers are degenerate cases of odd molecules.

Table 2 summarizes all the reactions that may be distinguished in the system. On the right to each reaction there are shown codes of molecules that react

Table 2. The list of reactions taking place during step growth alternating copolymerization of two bifunctional monomers

I	$ARA + BR'B \xrightarrow{4k_{00}} A\cdots B$		$\{1,0\} + \{0,1\} \rightarrow \{1,1\}$
II	$A\cdots B + ARA \xrightarrow{2k_{01}} A\cdots A$		$\{k,k\} + \{1,0\} \rightarrow \{k+1,k\}$
III	$A\cdots B + BR'B \xrightarrow{2k_{10}} B\cdots B$		$\{k,k\} + \{0,1\} \rightarrow \{k,k+1\}$
IV	$A\cdots A + BR'B \xrightarrow{4k_{10}} A\cdots B$		$\{k+1,k\} + \{0,1\} \rightarrow \{k+1, k+1\}$
V	$B\cdots B + ARA \xrightarrow{4k_{01}} A\cdots B$		$\{k,k+1\} + \{1,0\} \rightarrow \{k+1, k+1\}$
VI	$A\cdots B + A\cdots B \xrightarrow{2k_{11}} A\cdots B$		$\{k-i, k-i\} + \{i,i\} \rightarrow \{k,k\}$
VII	$A\cdots B + A\cdots A \xrightarrow{2k_{11}} A\cdots A$		$\{k-i, k-i\} + \{\{i+1,i\} \rightarrow \{k+1,k\}$
VIII	$A\cdots B + B\cdots B \xrightarrow{2k_{11}} B\cdots B$		$\{k-i, k-i\} + \{i,i+1\} \rightarrow \{k,k+1\}$
IX	$A\cdots A + B\cdots B \xrightarrow{4k_{11}} A\cdots B$		$\{k-i, k-i-1\} + \{i,i+1\} \rightarrow \{k,k\}$

and that are formed. The codes are simply the pairs of numbers of units RA_2 and $R'B_2$ in each molecule, respectively. This is not the only way of coding molecules for the purpose of kinetics analysis. In fact the method of coding depends on the kind of information sought for in the analysis. Cheng and Chiu [17] were able to obtain detailed information on reacting molecules in a multistage reaction system by using codes in the form of matrices or multicomponent vectors.

Kinetics equations are then written for the rates of change of concentrations of k,k-mers (k,k-1 mers and k-1,k-mer). Equation (38) gives just the equation for molecules of even degree of polymerization (k,k-mers). The numbers of the corresponding reactions in Table 2 are shown over each term.

$$\frac{dc_1}{dt} = -\kappa c \frac{dc_{k,k}}{dt} = \overset{(IV)}{4k_{01}c_{1,0}c_{k-1,k}} + \overset{(V)}{4k_{10}c_{0,1}c_{k,k-1}} + 2k_{11} \left[\overset{(VI)}{\frac{1}{2}\sum_{j=1}^{k-1} c_{j,j}c_{k-j,k-j}} + \overset{(IX)}{2\sum_{j=2}^{k-1} c_{j-1,j}c_{k-j+1,k-j}} \right.$$

$$\overset{(II)}{-2k_{01}c_{1,0}c_{k,k}} \overset{(III)}{- 2k_{10}c_{0,1}c_{k,k}} - 2k_{11}c_{k,k} \left. \left[\overset{(VI)}{\sum_{j=1}^{\infty} c_{j,j}} + \overset{(VII)}{\sum_{j=2}^{\infty} c_{j,j-1}} + \overset{(VIII)}{\sum_{j=2}^{\infty} c_{j-1,j}} \right] \tilde{H}_0 \right]_1 \quad (38)$$

The specific rate of dimer formation ($k=1$), in reaction I in Table 2, has to be considered separately. It reads

$$\left(\frac{dc_{1,1}}{dt}\right)_{form.} = 4k_{00}c_{1,0}c_{0,1} \quad (39)$$

Before we present some analytical solutions for simplified cases let us consider the general case where monomers react with substitution effect. (An example of such a monomer might be, say, 2,6-toluilenediisocyanate). In reactions I–IX there are four rate constants. We can define contributions to rate constants from the units, thus

$$k_A = \frac{k_{01}}{k_{11}} \quad \text{and} \quad k_B = \frac{k_{10}}{k_{11}} \quad (40)$$

The physical meaning of these constants is the following: k_A (or k_B) describes the reactivity of a functional group of the monomer relative to that of a group on a unit at the end of a chain.

Furthermore, we can assume that the contributions to the activation energy of reaction from the substitution degree of reacting units are additive. Thus, the activation energy for the reaction between i substituted unit A and j substituted unit B ($i,j=0,1$) is

$$\Delta E_{i,j} = \Delta E_{0,0} + i\psi_A + j\psi_B \tag{41}$$

Accordingly, the ratio of rate constants can be written as the contribution product:

$$\frac{k_{00}}{k_{11}} = k_A k_B \tag{42}$$

It is now only a matter of algebra to arrive at the following (master) set of kinetics equations [16]

$$\frac{\partial A}{\partial t} = \tilde{A}E - \tilde{A}\left(E_0 + 2\tilde{B}_0\right) \tag{43}$$

$$\frac{\partial B}{\partial t} = \tilde{B}E - \tilde{B}\left(E_0 + 2\tilde{A}_0\right) \tag{44}$$

$$\frac{\partial E}{\partial t} = \tfrac{1}{2}E^2 + 2\tilde{A}\tilde{B} - E\left(E_0 + \tilde{A}_0 + \tilde{B}_0\right) \tag{45}$$

The functions in Eqs. (43)–(45) are defined thus:

$$\tilde{A}(t,x,y) = k_A c_{1,0} x + \sum_{j=2}^{\infty} c_{j,j-1} x^j y^{j-1} \tag{46}$$

$$\tilde{B}(t,x,y) = k_B c_{0,1} y + \sum_{j=2}^{\infty} c_{j-1,j} x^{j-1} y^j \tag{47}$$

$$E(t,x,y) = \sum_{j=1}^{\infty} c_{j,j} x^j y^j \tag{48}$$

and the lack of a tilde over symbols A and B signifies that k_A and k_B have been set to 1. (They become functions identical with those in Eqs. (53) and (54), which follow). The subscript '0' denotes the value of a function at $x = y = 1$. It is not difficult to solve the set of Eqs. (43)–(45) numerically for any k_A and k_B.

A somewhat simpler set of equations can be obtained for the reaction of monomers where reactivities of functional groups remain unchanged. When we set

$$k_{00} = k_{10} = k_{01} = k_{11} = \tfrac{1}{2} \tag{49}$$

the system of kinetics equations of the type of Eq. (38) reduces to the following three equations describing the rates at which molecules appear or disappear:

$$\frac{dc_{k,k-1}}{dt} = \sum_{j=1}^{k-1} c_{j,j} c_{k-j,k-j-1} - c_{k,k-1} \left[\sum_{j=1}^{\infty} c_{j,j} + 2\sum_{j=1}^{\infty} c_{j-1,j} \right] \qquad (50)$$

$$\frac{dc_{k-1,k}}{dt} = \sum_{j=1}^{k-1} c_{j,j} c_{k-j-1,k-j} - c_{k-1,k} \left[\sum_{j=1}^{\infty} c_{j,j} + 2\sum_{j=1}^{\infty} c_{j,j-1} \right] \qquad (51)$$

and

$$\frac{dc_{k,k}}{dt} = \frac{1}{2}\sum_{j=1}^{k-1} c_{j,j} c_{k-j,k-j} + 2\sum_{j=1}^{k} c_{j-1,j} c_{k-j+1,k-j} - c_{k,k} \left[\sum_{j=1}^{\infty} c_{j,j} + \sum_{j=1}^{\infty} c_{j,j-1} + \sum_{j=1}^{\infty} c_{j-1,j} \right] \qquad (52)$$

including $c_{1,0}$ and $c_{0,1}$ since monomers can now be regarded as molecules A...A, and B...B. Equations (50)–(52) can be converted into a more manageable form by using the 'untilded' functions A and B, mentioned previously.

$$A(t,x,y) = \sum_{j=1}^{\infty} c_{j,j-1} x^j y^{j-1} \qquad (53)$$

$$B(t,x,y) = \sum_{j=1}^{\infty} c_{j-1,j} x^{j-1} y^j \qquad (54)$$

to obtain the set of differential equations equivalent to Eqs. (43)–(45), but with all tildes removed.

The moments of size distribution are given by

$$M_r(t) = \sum_{k=1}^{\infty} (2k-1)^r c_{k,k-1} + \sum_{k=1}^{\infty} (2k-1)^r c_{k-1,k} + \sum_{k=1}^{\infty} (2k)^r c_{k,k} \qquad (55)$$

The terms required to calculate these quantities are readily available from the sets of ordinary differential equations obtained by successive differentiation of Eqs. (43)–(45) with respect to x and y, followed by setting $x=y=1$. In fact, enough information is provided by differentiation of A, B, and E with respect to x, alone. Following the definitions of moments and averages, one can derive appropriate identities that eliminate the need to calculate other derivatives. Thus, in order to calculate number and weight average degrees of polymerization, it is sufficient to solve (usually numerically) the following sets of differential equations

$$\frac{dA_0}{dt} = -2A_0B_0; \quad \frac{dB_0}{dt} = -2A_0B_0; \quad \frac{dE_0}{dt} = 2A_0B_0 - E_0\left(\tfrac{1}{2}E_0 + A_0 + B_0\right) \quad (56)$$

$$\frac{dA_x}{dt} = E_xA_0 - 2A_xB_0; \quad \frac{dB_x}{dt} = E_xB_0 - 2B_xA_0;$$

$$\frac{dE_x}{dt} = 2A_xB_0 + 2B_xA_0 - E_x(A_0 + B_0) \quad (57)$$

and

$$\frac{dA_{xx}}{dt} = E_{xx}A_0 + 2E_xA_x - 2A_{xx}B_0; \quad \frac{dB_{xx}}{dt} = E_{xx}B_0 + 2E_xB_x - 2B_{xx}A_0;$$

$$\frac{dE_{xx}}{dt} = E_x^2 + 2A_{xx}B_0 + 4A_xB_x + 2B_{xx}A_0 - E_{xx}(A_0 + B_0) \quad (58)$$

(The subscript indicate the differentiation variables and '0' no differentiation before x and y are set to 1.) It is not difficult to solve the sets analytically for the stoichiometric mixture ($A_0 = B_0 = 1/2$). The solutions together with explicit distribution functions are collected in Table 3.

It has been shown [16] that the functions in Table 3 lead, as they should, to the same simple expressions for the number and weight average degrees of polymerization as those in Table 1.

Table 3. Explicit solutions of the sets of equations (Eqs. 56–58) and the distribution functions of polymer species for the step growth alternating copolymerization of two bi-functional monomers [16] with the stoichiometric monomer mixture, $c_{1,0}(0) = c_{0,1}(0) = \tfrac{1}{2}$

$A(t,x,y) = \dfrac{2x}{(2+t)^2 - t^2xy}$	$B(t,x,y) = \dfrac{2y}{(2+t)^2 - t^2xy}$	$E(t,x,y) = \dfrac{4txy}{(2+t)\left[(2+t)^2 - t^2xy\right]}$
$A_0(t) = \dfrac{1}{2(1+t)}$	$B_0(t) = \dfrac{1}{2(1+t)}$	$E_0(t) = \dfrac{t}{(1+t)(2+t)}$
$A_x(t) = \dfrac{(2+t)^2}{8(1+t)^2}$	$B_x(t) = \dfrac{t^2}{8(1+t)^2}$	$E_x(t) = \dfrac{t(2+t)}{4(1+t)^2}$
$A_{xx}(t) = \dfrac{t^2(2+t)^2}{16(1+t)^3}$	$B_{xx}(t) = \dfrac{t^4}{16(1+t)^3}$	$E_{xx}(t) = \dfrac{t^3(2+t)}{8(1+t)^3}$
$c_{k,k-1} = c_{k-1,k} = \dfrac{2t^{2(k-1)}}{(2+t)^{2k}}$	$c_{k,k} = \dfrac{4t^{2k-1}}{(2+t)^{2k+1}}$	

6
Random Homopolymerization of a Monomer with More Than Two Functional Groups

In this section we deal with the model step growth polymerization processes involving monomers having more than two functional groups. The Flory-Stockmayer model of homopolymerization of an f-functional monomer [18, 19] is included (see Table 4). This model is fundamental for understanding gelation processes [20]. Other processes relevant to polymer science which can be analyzed by using the Smoluchowski equation (Eq. 3) are also briefly described. All are random polymerization processes, i.e., functional groups react at random without any substitution effect and obey the Flory assumption of the independence of functional group reactivity on the size of species to which they are connected.

The Smoluchowski coagulation equation has effectively been applied to model random irreversible homopolymerization of the monomer types presented in Table 4. As can be seen, in all cases the coagulation kernel has the bilinear form:

$$K_{i,j} = A + B(i+j) + Cij \qquad (59)$$

Spouge [21] has shown that for any kernel of this form the Smoluchowski equation can be solved analytically, i.e., the distribution of molecular size of polymer molecules at any extent of reaction prior to the gel point (if any) or at least moments of the distribution can be expressed explicitly.

Of particular interest are the special cases with the product kernel

$$K_{i,j} = s_i s_j \qquad (60)$$

where

$$s_i = \alpha i + \beta \qquad (61)$$

Table 4. The types of monomers undergoing homopolymerization and corresponding bilinear kernels in the Smoluchowski coagulation equation [21]. The functional groups A react with each other in models 1 and 2 and with groups B only in models 3 through 5

Model	Type of Monomer	Coagulation Kernel, K_{ij}
1	RA_f	$[(f-2)i+2][f-2]j+2]$
2	RA_∞	ij
3	ARB_{f-1}	$(f-2)(i+j)+2$
4	ARB_∞	$i+j$
5	$A_g RB_{f-g}$	$2(g-2)(f-g-1)ij + (f-2)(i+j)+2$

and α and β are arbitrary real or complex numbers. Application of the Smoluchowski equation to models 1 and 2 belonging to this class have been reviewed several times [7, 22, 23]. Furthermore, it has been shown [24] that models with $K_{i,j}=i+j$ (model 4 in Table 4) can be converted into model 2. The general solution for $K_{i,j}=A+B(i+j)$ was provided by Drake and Write [25] and Treat [26]. Model 3 was first studied by Flory [9] who used statistical arguments. It describes the system leading to highly branched structures which can be consider as an one-step method of manufacturing dendrimers. Model 5 seems not to be relevant to polymer science and the reader is referred to the original paper by Spouge [27].

The Smoluchowski coagulation equation with the kernel given by Eq. (60)

$$\frac{dc_k}{dt} = \frac{1}{2}\sum_{i=1}^{k-1} s_i c_i s_{k-i} c_{k-i} - s_k c_k \sum_{j=1}^{\infty} s_j c_j \tag{62}$$

can be transformed into the following partial differential equation

$$\frac{\partial H}{\partial t} = \frac{x^{2-\beta}}{2}\left(\frac{\partial H}{\partial x}\right)^2 - x\left(\frac{\partial H}{\partial x}\right) H_x \tag{63}$$

with the function

$$H(t,x) = \sum_{i=1}^{\infty} c_i x^{s_i} \tag{64}$$

and

$$H_x(t) \equiv \left(\frac{\partial H}{\partial x}\right)_{x=1} = \sum_{i=1}^{\infty} s_i c_i \tag{65}$$

Alternatively, by using the function [23]

$$G(t,x) = \sum_{i=1}^{\infty} c_i e^{s_i x} \tag{66}$$

one arrives at

$$\frac{\partial G}{\partial t} = \frac{e^{-\beta x}}{2}\left(\frac{\partial G}{\partial x}\right)^2 - \frac{\partial G}{\partial x} G_x \tag{67}$$

with

$$G_x(t) = \sum_{i=1}^{\infty} s_i c_i = \frac{\partial G}{\partial x}\bigg|_{x=0} \tag{68}$$

As in previous sections, Eq. (63) is derived from Eq. (62) by multiplying it by x^{s_k} followed by summing over all k's. Use is made of the identities

$$s_k = s_i + s_{k-i} - \beta \tag{69}$$

and

$$\sum_{k=1}^{\infty} s_k c_k x^{s_k} = x \frac{\partial H}{\partial x} \tag{70}$$

For $\beta \neq 0$, differentiation of Eq. (61) with respect to x followed by putting $x=1$ yields

$$\frac{dH_x}{dt} = -\frac{\beta}{2} H_x^2 \tag{71}$$

Hence,

$$H_x = \frac{2}{\beta t + 2/(\alpha+\beta)} \tag{72}$$

For $\beta=0$, the result is $H_x = \alpha$.

The zeroth moment of the distribution is now obtained from Eq. (63) by setting $x=1$ to get

$$\frac{dH_0}{dt} = -\frac{1}{2} H_x^2 \tag{73}$$

and, upon solving

$$H_0 = \frac{\frac{2}{\alpha+\beta} - \alpha t}{\frac{2}{\alpha+\beta} + \beta t} \tag{74}$$

or, for $\beta=0$,

$$H_0 = 1 - \alpha^2 t/2 \tag{75}$$

Equation (63) can be solved by using the method of characteristics [6]. For the Flory-Stockmayer model where $s_i = (f-2)i + 2$ is the number of unreacted functionalities in an i-mer the solution is [28]

$$H(t,x) = \xi^f - \frac{f}{2} \xi^{2(f-1)} \frac{ft}{1+ft} \tag{76}$$

with the parameter ξ related to x and t through

$$x = \xi + ft\left(\xi - \xi^{f-1}\right) \tag{77}$$

(The solution is obtained by solving the set of three ordinary differential equations: $dx/dt = -xH_x - X$; $dX/dt = -XH_x$; and $dH/dt = -X^2/2$, where $X=\partial H/\partial x$, with the initial conditions $x(0) = \xi$; and $H(0) = \xi^f$. $H_x = f/(1+ft)$, as follows from Eq. (72) after substituting $\alpha = f - 2$ and $\beta = 2$.) The explicit form of the distribution of i-mers can be obtained from Eqs. (76) and (77) by the Lagrange expansion. The resulting distribution function for the Flory-Stockmayer model (model 1) is presented in Table 5.

The aggregation processes modeled by the Smoluchowski equation with the kernel $K_{i,j}=ij$ (model 2) or, more generally with $K_{i,j}=(ij)^\omega$ are relevant to many branches of physics [29–33]. The exponent ω is a geometric factor reflecting the mechanism of aggregation. If the rate of aggregation is taken to be proportional to the surface area of clusters, then ω reflects the dependence of the surface area on dimensionality, d. For compact clusters $\omega=1-1/d$. For aggregation or polymerization processes where steric hindrances become effective or intramolecular bonding may not be neglected, the factor ω should be taken as smaller than 1. Unfortunately, explicit solutions of the Smoluchowski equation are unknown for the cases with $0 < \omega < 1$. For the case with $\omega=1$ (model 2 in Table 4) the clusters aggregate with the rate proportion al to their volume (or mass). This case is also similar to the random evolution of graphs, the so called Erdös-Renýi process [34]. Both produce the same size distribution of components (aggregates or connected subgraphs). The differences between the two models have recently been discussed [35].

To solve Eq. (63) with $\alpha=1$ and $\beta=0$ ($K_{i,j}=ij$, model 2) one may again apply the method of characteristics to obtain

$$H(t,x) = \xi - \tfrac{1}{2}\xi^2 t \tag{78}$$

with

$$x = \xi \exp\left[(1-\xi)t\right] \tag{79}$$

Table 5. Moments and explicit distributions generated by the Smoluchowski coagulation equation (prior to gel point) for models 1 and 2 in Table 4. [For both models $M_1=1$]

Model	M_0	M_2	c_k
1	$\dfrac{1-\tfrac{1}{2}(f-2)ft}{1+ft}$	$\dfrac{1+2ft}{1-(f-2)ft}$	$\dfrac{f(fk-k)!}{k!(fk-2k+2)!}\dfrac{(ft)^{k-1}}{(1+ft)^{(f-1)k+1}}$
2	$1-t/2$	$\dfrac{1}{1-t}$	$c_k(t) = \dfrac{k^{k-2}}{k!}t^{k-1}e^{-kt}$

(Now, the set of differential equations is $dx/dt = -x^2X + x$; $dX/dt = xX^2 - X$; and $dH/dt = -x^2X^2/2$ with the initial conditions $x(0) = \xi$; $X(0) = 1$; and $H(0) = \xi$. $H_x = 1$.) The result of the Lagrange expansion of Eqs. (78) and (79) is presented in Table 5. As can be seen, the second moments (as well as higher ones) diverge at certain t. This divergency corresponds to gelation. The post-gelation relationships (including the moment expressions more general than those in Table 5) are discussed in Sect. 8, below.

7
The Equivalence Between Kinetics and Simple Statistical Models

The Smoluchowski coagulation equation deals with irreversible aggregation processes. There is no term in the equation describing backward reactions. Nevertheless, in 1943, Stockmayer [36] showed that for homopolymerization of an f-functional monomer the distribution of polymer species generated by the Smoluchowski equation is identical with that calculated by using statistical arguments. The statistical approach involves generation of the (most probable) distribution from the probabilities of functional groups being reacted. According to Whittle [37, 38], the statistically generated distribution is that in which there exists both local and global equilibrium between unreacted functional groups and those reacted to form bonds. As Spouge [21] has shown, for a distribution to be the equilibrium one the following combinatorial identity must hold:

$$(k-1)\frac{w_k}{k!} = \frac{1}{2}\sum_{i=1}^{k-1} K_{i,k-i} \frac{w_i}{i!} \frac{w_{k-i}}{(k-i)!} \qquad (80)$$

In Eq. (80) w_k is the number of ways of assembling a k-mer from distinguishable units. The identity simply states that the rate of assembling a k-mer from components i and $k-i$ (possible in $\binom{k}{i} = \frac{k!}{i!(k-i)!}$ ways) equals the rate of breaking one of $k-1$ bonds in the k-mer. For this identity to be valid, the kernel $K_{i,j}$ must be equal to the number of ways the components can be linked together. All kernels in Table 4 meet this requirement. In model 1 (Flory-Stockmayer), it is the number of ways the functional groups can react; in model 2, it is the number of units in the reacting components between which a link may be formed, etc. Spouge has also generalized this requirement by proving [39] that the identity at Eq. (80) is valid *if and only if* $K_{i,j}$ has the bilinear form given by Eq. (59). The case with a constant value of $K_{i,j}$, dealt with in Sect. 3, also belongs to the same class of kernels.

The existence of w_k related to the distribution of aggregate sizes and fulfilling the identity at Eq. (80) implies that the aggregation process leads to an equilibrium distribution. For the Flory-Stockmayer model (model 1 in Table 4) the number of assembling an acyclic k-mer from distinguishable units is [40]

$$w_k = \frac{f^k(fk-k)!}{[(f-2)k+2]!} \tag{81}$$

while for $K_{i,j}=ij$

$$w_k = k^{k-2} \tag{82}$$

It is now easy to see that if these w_k fulfill Eq. (80), the explicit distributions c_k in Table 5 necessarily fulfill the relation

$$(k-1)c_k = \frac{1}{2}\sum_{i=1}^{k-1} K_{i,k-i} c_i c_{k-i} \tag{83}$$

The same recurrence equation is obtained by solving successive terms of the Smoluchowski equation with $\sum_{i=1}^{\infty} s_i c_i = H_x$ equal to $f/(1 + ft)$ and 1 for models 1 and 2, respectively.

In the past, the equivalence between the size distribution generated by the Smoluchowski equation and simple statistical methods [9, 12, 40–42] was a source of some confusion. The Spouge proof and the numerical results obtained for the kinetics models with more complex aggregation physics, e.g., with a presence of substitution effects [43, 44], revealed the non-equivalence of kinetics and statistical models of polymerization processes. More elaborated statistical models, however, with the complete analysis made repeatedly at small time intervals have been shown to produce polymer size distributions equivalent to those generated kinetically [45]. Recently, Faliagas [46] has demonstrated that the kinetics and statistical models which are both the mean-field models can be considered as special cases of a general stochastic Markov process.

8
Gelling Systems with Substitution Effect

Modeling of a polymerization system becomes more and more complicated as the number of different kinds of monomers and reactive groups increases. To the complexity add substitution effects that are quite common for monomers of practical importance. Furthermore, the substitution effects may be considered as a method of relaxing the Flory postulate of the independence of reactivity of the size of molecules. They model conversion dependence of the reactivity in the system.

For simplicity, we start with a single tri-functional monomer RA_3, the functional groups of which react with each other with the first shell substitution effect. It is now convenient to write down six types of reactions [44], one for each pair of reacting units of substitution degree 0, 1, and 2. They are presented in Table 6 together with the appropriate rate constants. Similarly as in previous sec-

Table 6. The types of reactions in model polymerization of RA3 monomer with functional groups reacting with the first shell substitution effect. The subscripts at the symbols of reagents correspond to the codes of molecules (i,j-mers, see the text) and those at the rate constants indicate the degrees of substitution of reacting units. The left and right column correspond to disappearance and formation of an i,j-mer, respectively

$P_{0,0} + P_{0,0} \xrightarrow{9k_{00}} P_{2,0}$	$P_{0,0} + P_{0,0} \xrightarrow{9k_{00}} P_{2,0}$
$P_{i,j} + P_{0,0} \xrightarrow{6k_{01}} P_{i,j+1}$	$P_{i,j-1} + P_{0,0} \xrightarrow{6k_{01}} P_{i,j}$
$P_{i,j} + P_{0,0} \xrightarrow{3k_{02}} P_{i+1,j-1}$	$P_{i-1,j+1} + P_{0,0} \xrightarrow{3k_{02}} P_{i,j}$
$P_{i,j} + P_{r,s} \xrightarrow{4k_{11}} P_{i+r,j+s+2}$	$P_{i-r+1,j-s-1} + P_{r+1,s-1} \xrightarrow{4k_{11}} P_{i,j}$
$P_{i,j} + P_{r,s} \xrightarrow{2k_{12}} P_{i+r-1,j+s}$	$P_{i-r+1,j-s} + P_{r,s} \xrightarrow{2k_{12}} P_{i,j}$
$P_{i,j} + P_{r,s} \xrightarrow{k_{22}} P_{i+r,j+3-2}$	$P_{i-r,j-s+1} + P_{r,s+1} \xrightarrow{k_{22}} P_{i,j}$

tions, the constant $k_{\alpha\beta}$ ($\alpha, \beta = 0, 1, 2$) describes reactivity of functional groups in a pair of reacting units of substitution degree α and β. All unreacted groups on a unit have the same reactivity which depends solely on its substitution degree.

Table 6 presents the reactants and products of reactions. The pairs of numbers in subscripts are codes for molecules. The first code number for an (i,j)-mer is the number of units in polymer molecule with exactly one reacted functional group (substitution degree 1) while the second is the number of units with two reacted groups (substitution degree 2). The symbol $P_{0,0}$ stands for the monomer. Note that there is no need to specify the number, say k, of units with all three groups reacted; they do not enter reactions anymore and, as long as the molecule has no cycles, the two parameters provide all information on its size n. The following two relations hold: the obvious one $n = i + j + k$ and $2(n - 1) = i + 2j + 3k$. The second relationship sums up all reacted groups that form the $n - 1$ bonds in an acyclic molecule (two groups per each bond). The parameter k may now be eliminated to yield

$$n = 2i + j - 2 \tag{84}$$

Kinetic equations describing the rates of formation and further reactions of (i,j)-mers are more complicated. They contain up to ten different terms. Considerable simplification and relatively easy access to average parameters of reacting systems (such as average degrees of polymerization) can be achieved by applying the assumption on the additivity of activation energies similar to that adopted in Sect. 3. The rate constants can then be split into the product form:

$$k_{\alpha\beta} = k^* k_\alpha k_\beta \tag{85}$$

($\alpha, \beta = 0, 1, 2$) where k^* stands for the substitution independent part of the constant. The number of rate constants can now be reduced to just two [44]:

$$\kappa_1 = \frac{2k_1}{3k_0} \; ; \; \kappa_2 = \frac{k_2}{3k_0} \tag{86}$$

With $9k^*k_0^2$ absorbed into time units, the relative rate constant with which a monomer reactive group reacts is simply unity, while the relative rate constant for mono- and disubstituted units is κ_1 and κ_2, respectively.

Note that for the random case, equivalent to the Flory-Stockmayer model (model 1 in Table 4), $\kappa_1 = 2/3$ and $\kappa_2 = 1/3$.

For the function defined as

$$H(t, x, y) = c_1(t) + \sum_{i=2}^{\infty} \sum_{j=0}^{\infty} c_{ij}(t) [\kappa_1 x]^i [\kappa_2 y]^j \tag{87}$$

($c_1(t) = c_{0,0}(t)$) the set of kinetics equations for the reactions in Table 6) can be transformed into the single partial differential Smoluchowski-like equation in the form

$$\frac{\partial H}{\partial t} = \frac{1}{2} \left[\kappa_1 x c_1 + \kappa_2 y \frac{\partial H}{\partial x} + \frac{\partial H}{\partial y} \right]^2 - (c_1 + H_x + H_y) \left[c_1 + \kappa_1 x \frac{\partial H}{\partial x} + \kappa_2 y \frac{\partial H}{\partial y} \right] \tag{88}$$

where

$$H_x = H_x(t) = \frac{\partial H}{\partial x}\bigg|_{x=1/\kappa_1, y=1/\kappa_2} = \kappa_1 \sum_{i=2}^{\infty} \sum_{j=0}^{\infty} i c_{i,j} \tag{89}$$

$$H_y = H_y(t) = \frac{\partial H}{\partial y}\bigg|_{x=1/\kappa_1, y=1/\kappa_2} = \kappa_2 \sum_{i=2}^{\infty} \sum_{j=0}^{\infty} j c_{i,j} \tag{90}$$

Equations (89) and (90) define functions of time obtained by differentiation followed by substitution of x and y by $1/\kappa_1$ and $1/\kappa_2$, respectively, so that the dummy variable terms cancel in the derivative.

As in the previous sections, Eq. (88) is obtained by multiplying by $(\kappa_1 x)^i (\kappa_2 y)^j$ all possible terms that can be written down for formation, and disappearance, of an i,j-mer followed by summing up the result over all i and j.

In order to verify Eq. (88) note that the term corresponding to the rate of formation of (i,j)-mer by reaction of two monosubstituted units (reaction 4 in Table 6) is

$$\frac{1}{2} \kappa_1^2 \sum_{r=2}^{i-1} \sum_{s=0}^{j-1} (i-r+1) c_{i-r+1, j-s-1} (r+1) c_{r+1, s-1} \tag{91}$$

(in parantheses are the numbers of functional groups available for reaction).

This term yields, after multiplying by $(\kappa_1 x)^i (\kappa_2 y)^j$ and summing up over i and j,

$$\frac{1}{2}\left[\kappa_2 y \frac{\partial H}{\partial x}\right]^2 \tag{92}$$

The disappearance term, say, the one corresponding to reaction 5 in Table 6 reads

$$-\kappa_1 \kappa_2 i c_{i,j} \sum_{r=2}^{\infty} \sum_{s=1}^{\infty} s c_{r,s} \tag{93}$$

and turns into

$$-\kappa_1 x \frac{\partial H}{\partial x} H_y \tag{94}$$

of Eq. (88).

It is not difficult to extend the reasoning to the general case of RA_f polymerization with the first shell substitution effect approximation. By defining the function

$$H(t, x_1, ..., x_{f-1}) = c_1(t) + \sum_{i=2}^{\infty} \sum_{j=0}^{\infty} .. \sum_{z=0}^{\infty} c_{ij..z}(t) [\kappa_1 x_1]^i [\kappa_2 x_2]^j ..[\kappa_{f-1} x_{f-1}]^z \tag{95}$$

where the κ_r stand for rate constants relative to that of monomer

$$\kappa_r = (f-r) k_r \Big/ f k_0 \quad ; r = 1, 2, .., f-1 \tag{96}$$

one arrives at the general Smoluchowski-like equation in the form

$$\frac{\partial H}{\partial t} = \frac{1}{2}\left[\kappa_1 x_1 c_1 + \kappa_2 x_2 \frac{\partial H}{\partial x_1} + .. + \kappa_{f-1} x_{f-1} \frac{\partial H}{\partial x_{f-2}} + \frac{\partial H}{\partial x_{f-1}}\right]^2$$

$$-(c_1 + H_1 + .. + H_{f-2} + H_{f-1})\left[c_1 + \kappa_1 x_1 \frac{\partial H}{\partial x_1} + .. + \kappa_{f-2} x_{f-2} \frac{\partial H}{\partial x_{f-2}} + \kappa_{f-1} x_{f-1} \frac{\partial H}{\partial x_{f-1}}\right] \tag{97}$$

with $H_r(t)$ defined by relations analogous to Eqs. (89) and (90). An equation equivalent to Eq. (97) was first derived by Kuchanov [43, 47]. If there is no substitution effect we have

$$\kappa_r = (f-r)/f \; ; (r = 1, 2, .., f-1) \tag{98}$$

and the system becomes the general Flory-Stockmayer one.

An interesting new model is provided by Eq. (97) with $\kappa_r=1$ for all r. In the limit of $f \to \infty$ it becomes the RA_∞ model, fully equivalent to that described by Eq. (3) with $K_{i,j}=ij$. At bounded f it describes the evolution of f-trees, i.e., a process similar to that dealt with by Erdös and Renýi [20, 34], but with substitution degree restriction imposed on the vertices (units) and cycle formation disallowed.

Explicit solutions of Eqs. (88) or (97) are not known, even for the Flory-Stockmayer model. The concentrations of individual species $c_{i,j}(t)$ for that model are available, but only by recursive solution of appropriate kinetics equations (before they are summed up to yield Eqs. (88) or (97)). The function $c_1(t) + H_x(t) + H_y(t)$ which is needed to solve the equations is

$$c_1 + H_x + H_y = \frac{3}{t+3} \tag{99}$$

Physically, it is the fraction of functional groups available to reaction (and hence equivalent to H_x in Eqs. (65) and (72)). Equation (99) reveals that for this version of the FS model the time units are „nine times longer" than those from direct application of the original Smoluchowski equation.

Equation (88) was used to demonstrate [44] the differences in distributions of molecular sizes in pre-gel stages of an RA_3 polymerization with substitution effects as calculated according to a statistical and a kinetics model. The moments of distribution and the gel points were calculated by the numerical solution of a few ordinary differential equations that were derived from Eq. (88) and compared with analogous quantities calculated from a statistical model.

To illustrate the application of Eq. (88) we will derive equations needed to calculate the second moment of molecular size distribution. The r-th moment of the size distribution which, taking into account Eq. (84), is defined thus:

$$M_r(t) = c_1(t) + \sum_{i=2}^{\infty} \sum_{j=0}^{\infty} (2i+j-2)^r c_{i,j}(t) \tag{100}$$

can be expressed as linear combinations of successive derivatives of H with respect to x and/or y of at most order r, calculated at $x=1/\kappa_1$ and $y=1/\kappa_2$.

By raising the term right after the sum signs in Eq. (100) to the second power () and regrouping the result we get

$$M_2 = c_1 + \sum_i \sum_j \left[4i(i-1) + 4ij + j(j-1) - 4i - 3j + 4\right]c_{ij} = \frac{4}{\kappa_1^2}H_{xx} + \frac{4}{\kappa_1\kappa_2}H_{xy} + \frac{1}{\kappa_2^2}H_{yy} - \frac{4}{\kappa_1}H_x - \frac{3}{\kappa_2}H_y + 4H_0 - 3c_1 \tag{101}$$

The subscripts of H indicate variables with respect to which differentiation has to be made in Eq. (87) (0 for no differentiation) before the dummy variables are set equal to $1/\kappa_1$ and $1/\kappa_2$ as in Eqs. (89) and (90). The required derivatives are obtained from the Smoluchowski-like equation (88) by

i. setting $x=1/\kappa_1$ and $y=1/\kappa_2$ to get (upper dot denotes time derivative)

$$\dot{H}_0 = -\tfrac{1}{2}(c_1 + H_x + H_y)^2 \qquad (102)$$

ii. setting $x=0$ and $y=0$ to get

$$\dot{c}_1 = -c_1(c_1 + H_x + H_y) \qquad (103)$$

iii. differentiating appropriately Eq. (88) with respect to dummy variables x and y and then setting $x=1/\kappa_1$ and $y=1/\kappa_2$ to get

$$\dot{H}_x = -k_1(H_x - c_1)(c_1 + H_x + H_y) \qquad (104a)$$

$$\dot{H}_y = -k_2(H_y - H_x)(c_1 + H_x + H_y) \qquad (104b)$$

$$\dot{H}_{xx} = (k_1 c_1 + H_{xx} + H_{xy})^2 - 2k_1 H_{xx}(c_1 + H_x + H_y) \qquad (104c)$$

$$\dot{H}_{xy} = (k_1 c_1 + H_{xx} + H_{xy})(k_2 H_x + H_{xy} + H_{yy}) - [(k_1 + k_2)H_{xy} - k_2 H_{xx}](c_1 + H_x + H_y) \qquad (104d)$$

$$\dot{H}_{yy} = (k_2 H_x + H_{xy} + H_{yy})^2 - 2k_2(H_{yy} - H_{xy})(c_1 + H_x + H_y) \qquad (104e)$$

Equations (102)–(104) are the Riccati differential equations that have no solutions in quadratures for arbitrary κ_1 and κ_2. They can, however, be easily solved numerically to obtain the desired second moment of the distribution (and hence the weight average degree of polymerization) as a function of time, according to Eq. (101).

Only for the FS model ($\kappa_1=2/3$, $\kappa_2=1/3$) have explicit solutions of Eqs. (102)–(104) been found. The functions H_{xx}, H_{xy}, H_{yy} have, as expected, a singularity at $t=3$ of the form $(3-t)^{-1}$. The singularity corresponds to the gel point. At $t_c=3$, where t_c is the time where the gel point is, these functions simultaneously diverge. No modifications of Eq. (88) that would be valid beyond this point are known.

For arbitrary κ_1 and κ_2 no explicit expression for the gel point exists. In statistical models, the gel point can be found as the time (or conversion) when the fraction of fully reacted units reaches one third of the fraction of monosubstituted units. Simply, when there is not enough terminating units to saturate all branches extending from those with all three groups reacted, the infinite (gel) molecule necessarily forms. Note that c_1, H_x/κ_1, and H_y/κ_2 are just the fractions of units with 0, 1 and 2 reacted functional groups, respectively. The fraction of those with all three groups reacted is $1 - c_1 - H_x/k_1 - H_y/k_2$. It is interesting that, except for the FS case, the simple condition for gelation deduced from statistical models does not hold for the kinetics one (Eq. 88). The functions H_{xx}, H_{xy}, H_{yy} are found to diverge before that condition was met.

Modeling with the Smoluchowski-like equation generalized to take into account FSSE is not limited to the simple RA_f polymerization. A kinetics approach similar to that described in this section have been used to study crosslinking reactions of epoxy resins with components introduced into the system at different times [17]. Kinetic equations analogous to Eq. (101) have been derived [48] for an $RA_2 + R'B_3$ system as well as for systems containing 3-functional monomers having functional groups of intrinsically different reactivities [49].

The effect of changes in reactivity of monomers due to substitution on the process of polymerization, mostly on the position of gel point was studied by using Monte Carlo calculations applying essentially the same approach as that presented in this section [50–52].

9
Post-Gelation Relationships

In principle, the Smoluchowski coagulation equation and its modifications provide the correct distributions of aggregate sizes at positive values of the parameter t. For some kernels the solutions are limited to the range of $0 \leq t \leq t_c$. At t_c the second and higher moments of distribution diverge. In this section the changes in distribution for $t > t_c$ are discussed.

It can be seen from Table 5 that for models 1 and 2 the gelation time is at

$$t_c = \frac{1}{f(f-2)} \quad \text{and} \quad t_c = 1 \qquad (105)$$

respectively. (Simply the denominator in M_2 is zero.) If we look at the explicit distributions in Table 5, we find that the concentrations for every $k > 1$ initially increase with time and then peaks and decays. The value of t_c is the limiting time where the molecule of size $k \to \infty$ reaches its maximum concentration

$$t_c = \lim_{k \to \infty} t_{max}(k) \qquad (106)$$

(t_{max} is obtained by setting $dc_k/dt=0$.) By that time, the concentrations of all finite (sol) molecules have already passed their peak values and decrease because of the growth of the largest species. The largest molecule can then be identified with the gel. Since $k \to \infty$, the expression for its concentration cannot be evaluated.

At the critical time when gel appears, t_c, the first moment of the distribution, which is constant for $t \leq t_c$ and reflects mass conservation in the system, starts to decay, corresponding to the loss of units to the gel molecule. The initial rate of the decay is easy to deduce. Since, for model 1, direct differentiation of the explicit form of concentration (Table 5) yields

$$\frac{dc_k}{dt} = c_k \frac{k-1-[(f-2)k+2]ft}{t(1+ft)} \tag{107}$$

the rate of decay of the first moment is

$$\left.\frac{dM_1}{dt}\right|_{t_c} = \sum_{k=1}^{\infty} k \left.\frac{dc_k}{dt}\right|_{t_c} = -\frac{f^2(f-2)}{f-1} \tag{108}$$

(More precisely the upper limit of summation in Eq. (108) should be set to some value L and allow to grow to infinity only after $t_c = 1/f(f-2)$ is substituted.) Similar reasoning for model 2 leads to the initial decay rate equal [7] to -1.

The Smoluchowski equation was thought of as failing beyond the gel point [22]. Dusek [53] and Ziff [54] have demonstrated this not to be true for the Flory-Stockmayer model. At the same time Leyvraz and Tschudi [32] presented an elegant and more general derivation of the solutions of the Smoluchowski equation both prior to and beyond the gel point.

To keep the mathematics as simple as possible, here we sketch the arguments of Ziff [54] and Ziff and Stell [28], who carefully analyzed the post-gelation behavior of model 1. Ziff and Stell distinguish three versions of its post-gel behavior.

The Flory model is the version where the equivalence between kinetics and statistical descriptions is extended to the post-gel stage of polymerization. Consequently, the functional groups are assumed to continue to react at random with no distinction on whether they belong to sol molecules or to gel. To analyze this version one can use the explicit form of function H. As usual, the moments are available through successive derivatives of H (Eq. 76) with respect to x calculated at $x=1$. We may rewrite Eq. (77) in the form

$$1 = \zeta + ft\left(\zeta - \zeta^{f-1}\right) \tag{109}$$

where ζ stands for ξ at $x=1$.

Equation (109) has two roots in the half-open range $(0,1]$. The trivial root, $\zeta = 1$, and the root given by the equation

$$\zeta + \zeta^2 + \ldots + \zeta^{f-2} = \frac{1}{ft} \tag{110}$$

obtained from Eq. (109) by removing the factor $1 - \zeta$. The second root, however, becomes accessible only past the gel point. Bearing that in mind we can now express the moments of the distribution by using Eq. (76) with the lowest positive root of Eq. (110). The equations will then apply to the whole system prior the gel point and to the sol fraction only beyond gelation. Thus,

$$M_0 = H(t,1) = \frac{1}{2}\left[\frac{f}{1+ft}\zeta^{f-1} - (f-2)\zeta^f\right] \tag{111}$$

(Substitute ζ^{f-1} from Eq. (109) into the second term of Eq. (76)),

$$M_1 = \frac{H_\xi/x_\xi\Big|_{x=1} - 2M_0}{f-2} = \zeta^f \qquad (112)$$

and

$$M_2 = \frac{1}{(f-2)^2}\left[\frac{H_{\xi\xi}x_\xi - H_\xi x_{\xi\xi}}{x_\xi^3}\bigg|_{x=1} - 3(f-2)M_1 - 2M_0\right] = \frac{2(1+ft)\zeta^f - \zeta^{f-1}}{1+ft-ft(f-1)\zeta^{f-2}} \qquad (113)$$

where the subscript ξ ($\xi\xi$) indicates that the functions in Eqs. (76) and (77) are differentiated once (twice) with respect to ξ,

The concentration of functional groups in the sol is

$$H_x = \frac{f}{1+ft}\zeta^{f-1} \qquad (114)$$

etc.

The reader familiar with the cascade theory will notice that the root $0 < \zeta \leq 1$ is related to the extinction probability, v, i.e., the probability of a unit chosen at random to belong to a sol molecule [14, 55]. This probability is [53]

$$v = \zeta^{f-1} \qquad (115)$$

The second version of the post-gel analysis is equivalent to that of Stockmayer [19]. He assumed that the distribution of sizes within sol past the gel point remains unchanged and the ratios of concentrations can still be deduced from those given in Table 5. Ziff and Stell [28] concluded that this assumption is equivalent to a disregard of reactions between gel and sol. Consequently, the concentration of each k-mer as well as that of the sol fraction decays according to the equation

$$c_k(t) = c_k(t_c)\frac{(f-1)t_c}{ft-t_c} \qquad (116)$$

Ziff and Stell [28] themselves proposed a third version of the post gel analysis which differed from Flory's one by disregarding the reactions taking place within gel molecule (no intramolecular reactions both in sol and gel).

The analysis for model 2 of Table 4 ($K_{i,j}=ij$) using the Flory model leads to the following moment expressions:

$$M_0 = \zeta - \zeta t/2; \quad M_1 = \zeta; \quad M_2 = \frac{\zeta}{1-\zeta t} \qquad (117)$$

where ζ is the real root of the equation

$$(1-\zeta)t+\log\zeta = 0 \tag{118}$$

Beyond the gel point ($t > 1$), the nontrivial root ($\zeta \ne 1$) should be taken into account.

Of interest also are the results of applications of the Smoluchowski equation for systems with more complex aggregation physics than that provided by bilinear kernels. Leyvraz and Tschudi [32] conjectured that for the kernel $K_{i,j}=(ij)^\omega$ gelation occurs only when $\omega > 1/2$. The post-gelation behavior of a general system with a multiplicative kernel given by Eq. (60) has been analyzed by van Dongen et al. [56]. By assuming that the distribution past the gel point could be expressed through that at $t=t_c$, thus

$$c_k(t) = \frac{c_k(t_c)}{1+b(t-t_c)} \tag{119}$$

with

$$b = \sum_k s_k c_k(t_c) \tag{120}$$

they were able to obtain the recurrence relation very similar to Eq. (83)

$$(s_k - 1)n_k = \frac{1}{2}\sum_{i+j=k} s_i s_j n_i n_j \tag{121}$$

where

$$(s_k - 1)n_k = \frac{1}{2}\sum_{i+j=k} s_i s_j n_i n_j \tag{122}$$

Post-gel distribution functions or their Laplace transforms were derived for several special cases including $s_k=k^\omega$. Unfortunately, no analytical method of predicting t_c has been found to exist. A numerically derived relationship between t_c and ω have recently been published [57]. With time units scaled so that $t_c=1$ for model 1 it reads

$$t_c = \frac{1}{8.902\cdot 10^{-2} - 1.3034\omega + 2.2144\omega^2} \tag{123}$$

10
Attempts at Taking Into Account Cyclization Reactions

The Smoluchowski coagulation equation describes the rate of formation of acyclic aggregates. Only then it describes the evolution of a Markovian distribution [34]. Strictly speaking the Smoluchowski equation simply disregards any cycle formation. For polymers this is true for models with high functionality

(RA_∞ or BRA_∞) where the 'wastage' of functionalities in cycle formation can be considered negligibly small. There have been attempts to include cyclization reactions into polymerization models, but, except perhaps the linear polymerization models, they are usually very crude or just not realistic.

The set of kinetics equations for linear polycondensation can be written as [58]

$$P_i + P_j \xrightarrow{\sigma k_{i,j}} P_{i+j}$$
$$P_i \xrightarrow{k_i^{(c)}} C_i \qquad (124)$$

where the first equation is the same as that in Eq. (7) and the second one describes the cyclization reactions. C_i stands for a cycle of size i. The rate constant in the second reaction is usually made dependent on the size of molecule. The standard assumption is [59] that it has the form

$$k_i^{(c)} \sim i^{-3/2} \qquad (125)$$

This assumption is equivalent to considering the polymer molecule as a Gaussian chain. For a Gaussian chain the probability of the two ends colliding in three-dimensional space is proportional to its length to the power –3/2. For the Kuhn (or freely-jointed chain) model the same assumption may be taken for sufficiently long chains [60]. For linear polymers in good solvents, no similar simple assumption can be adopted. To study cyclization one has to resort to more sophisticated mathematical treatments (see, e.g. [61]).

Since in irreversible linear polymerization the cyclic molecules once formed do not react further, the kinetics equations can be written thus (here C_i is the concentration of cycles):

$$\frac{dc_k}{dt} = \frac{1}{2}\sum_{i=1}^{k-1} c_i c_{k-i} - c_k \sum_{i=1}^{\infty} c_i - k_k^{(c)} c_k$$
$$\frac{dC_k}{dt} = k_k^{(c)} c_k \qquad (126)$$

A kinetics approach to model polymerizations of RA_∞ and RA_f monomers including cyclization was used by Lu and Bak [62]. They formulated and solved four systems of equations, in which cyclization terms were added to the Smoluchowski equation:

$$\frac{dc_k}{dt} = \frac{1}{2}\sum_{i=1}^{k-1} K_{i,k-i} c_i c_{k-i} - c_k \sum_{i=1}^{\infty} K_{i,k} c_i - \lambda c_k \qquad (127)$$

with the parameter λ either a constant or linear function of the size of k-mers (taken as λs_k with $s_k = \alpha k + \beta$). The disadvantage of the model is that it means that the species that undergo cyclization are automatically removed from considerations. An explicit gel time was derived for the kernel $K_{i,j} = ij$ and size independent λ. It reads (for the monodisperse initial conditions):

$$t_c = -\frac{1}{\lambda}\log(1-\lambda) = 1 + \frac{\lambda}{2} + \frac{\lambda^2}{3} + .. + \frac{\lambda^i}{i+1} + ..\quad (128)$$

Equation (128) can be derived by converting Eq. (127) into the partial differential equation of the same type as Eq. (63) and solving the ordinary differential equation for the second moment

$$\frac{dM_2}{dt} = M_2^2 - \lambda M_2 \quad (129)$$

Note, however, that for constant λ, it is the monomer which at small t contributes to cyclization the most, which is physically unrealistic.

It is not very difficult to extend formally the treatment presented in Sect. 8, namely the Smoluchowski-like equation (Eq. 88), to model, besides the substitution effect, the ability of functional groups to react intramolecularly. For the simplest case of RA_3 homopolymerization, a crude method [63] is to code the molecules with four indices two of which count the units with two or three reacted functional groups that are engaged in cycles. The Smoluchowski-like equation reads then

$$\frac{\partial H}{\partial t} = \frac{1}{2}\left[\kappa_1 u H_0 + \kappa_2 x \frac{\partial H}{\partial u} + y \frac{\partial H}{\partial v} + \frac{\partial H}{\partial x}\right]^2 -$$
$$(H_0 + H_u + H_v + H_x)\left[H_0 + \kappa_1 u \frac{\partial H}{\partial u} + \kappa_2 v \frac{\partial H}{\partial v} + \kappa_2 x \frac{\partial H}{\partial x}\right] +$$
$$\frac{1}{2}\left[\left[(\kappa_2 v)^2 - (\kappa_1 u)^2\right]\lambda_1 \frac{\partial^2 H}{\partial u^2} + 2\left[y\kappa_2 v - \kappa_1 u \kappa_2 x\right]\lambda_2 \frac{\partial^2 H}{\partial u \partial x} + \left[y^2 - (\kappa_2 x)^2\right]\lambda_3 \frac{\partial^2 H}{\partial x^2}\right] \quad (130)$$

and describe the time evolution of the generating function

$$H(t,u,v,x,y) = H_0(t) + \sum_i \sum_j \sum_l \sum_m C_{i,j,l,m}(t)[\kappa_1 u]^i [\kappa_2 v]^j [\kappa_2 x]^l [y]^m \quad (131)$$

where

$$H_\alpha(t) = \frac{\partial H}{\partial \alpha}\bigg|_{u=1/\kappa_1, v=1/\kappa_2, x=1/\kappa_2, y=1} \quad (\alpha = u,v,x,y) \quad (132)$$

and u, v, x and y are the auxiliary variables labelling the types of units (see Fig. 2). As many as three parameters λ_i ($i=1,2,3$) model the extent of intramolecular reaction, each for a pair of differently substituted units (they may be considered functions of time, size of molecules, dilution of the system, etc.). In one approach they were assumed constant and estimated by using graph theoretical methods

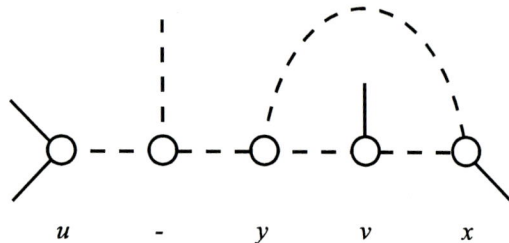

Fig. 2. A single molecule formed in RA$_3$ polymerization. The molecule contains various types of units. The labels of units are the same as the dummy variables in Eq. (130): u – monosubstituted unit with two unreacted groups (*solid lines*), v – disubstituted unit, x – disubstituted unit engaged in cycle, y – trisubstituted unit engaged in cycle

[64] to be in the ratio $\lambda_3/\lambda_2 \approx 1{,}5$ and $\lambda_2/\lambda_1 \approx 2$. Unfortunately, Eq. (130) has not been solved or examined numerically, as yet.

11
Other Remarks on the Smoluchowski Coagulation Equation

The continuous form of the Smoluchowski coagulation equation

$$\frac{\partial c(x,t)}{\partial t} = \frac{1}{2}\int_0^x K(x-u,u)c(x-u,t)c(u,t)du - c(x,t)\int_0^\infty K(x,u)c(u,t)du \qquad (133)$$

was first given by Müller [65]. Here, x and u are real numbers standing for aggregate sizes and $K(x,u)$ is a collision frequency function. Another method of modeling aggregation processes is the time-dependent Markovian process known as the Lushnikov process [66]. The analysis by Faliagas [46] is in fact an elegant application of Lushnikov's model to f-functional polymerization.

Considerable attention has been devoted to aggregation processes, the physics of which is well described by the Smoluchowski equation with so-called homogeneous kernels (i.e., with $K_{\alpha i,\alpha j}=\alpha^\lambda K_{i,j}$). Here, parameter λ is the degree of homogeneity. To describe the size distribution of aggregates in a finite system, functions have been proposed [67] in the form

$$c_k(t) \sim \frac{1}{[s(t)]^\theta} f\left(\frac{k}{s(t)}\right) \qquad (134)$$

where $f(\ )$ is a scaling function, or, more specifically, as

$$c_k(t) \sim \frac{1}{[s(t)]^\theta}\left[\frac{s(t)}{k}\right]^\tau \exp\left(-\frac{k}{s(t)}\right) \qquad (135)$$

where $s(t)$ is a scaling parameter usually identified with the z-average size of aggregates, which behaves according to

$$s(t) \sim \begin{cases} t^z; & \lambda < 1 \\ e^t; & \lambda = 1 \\ (t_{cr} - t)^{z'}; & \lambda > 1 \end{cases} \quad (136)$$

The scaling function (Eq. 135) holds for both time t and aggregate size k sufficiently large [68]. For particular kernels, the exponents, z and z', have been calculated. They are summarized in Table 7.

As mentioned previously, the most relevant to polymer science seems to be the product homogeneous kernel $K_{i,j}=(ij)^\omega$, (the homogeneity degree $\lambda=2\omega$). According to van Dongen and Ernst [69], exponents in the scaling relation given by Eq. (135) are

$$\tau = \omega + \frac{3}{2} \quad (137)$$

and

$$z' = \frac{1}{\frac{1}{2} - \omega} \quad (138)$$

They also pointed out that the scaling function at the gel point reduces to (cf. [31]).

$$c_k(t_c) \sim k^{-\tau} \quad (139)$$

The weight average degree of polymerization changes in the vicinity of gel point according to [70]

$$M_w \sim (t_{cr} - t)^{-\gamma} \quad (140)$$

with

$$\gamma = \frac{3 - 2\omega}{2\omega - 1}. \quad (141)$$

For negative ω, the coagulation leads after long time to a log normal distribution [71].

Table 7. The exponents in the function (Eq. 135) describing the asymptotic distribution of aggregates in a process modeled with the Smoluchowski coagulation equation

$K_{ij} =$	const.	$i+j$	ij
θ	2	2	2.5
t	0	1.5	2.5
z	1	∞	–
z'	–	–	–1

Critical times at gelation and exponents θ in Eq. (135) for several values of ω in the range 0.55–1 have recently been calculated numerically by Krivitsky [72].

On the other hand, in the Erdös-Renýi process, the size of a „typical" largest component in a system comprising N units has been proved to be proportional to $\log N$ up to the gel point. Beyond the gel point it (i.e., gel molecule) has a size proportional to N itself [20, 34].

An interesting modification of the Smoluchowski equation leading to the so called generalized Smoluchowski equation [73] can be obtained by adding the terms describing the rates at which clusters or polymer molecules undergo fragmentation:

$$\frac{dc_k}{dt} = \frac{1}{2}\sum_{i=1}^{k-1} K_{i,k-i} c_i c_{k-i} - c_k \sum_{i=1}^{\infty} K_{i,k} c_i - \frac{1}{2}\sum_{i=1}^{k-1} \overline{K}_{i,k-i} c_k + \sum_{i=1}^{\infty} \overline{K}_{i,k} c_{i+k} \qquad (142)$$

is the rate constant of the splitting of an $(i+j)$-mer into i- and j-mers. The third sum is the rate at which k-mer dissociate into smaller species, while the fourth is the rate at which it is formed from bigger ones. In general, the fragmentation terms do not complicate seriously the mathematical treatment of combined aggregation-fragmentation models [74–76].

Acknowledgments. This work was financially supported by the Polish Committee of Scientific Research, grant no. 3 T09A 038 011. We are grateful to Professors K. Dušek of Prague and R.F.T. Stepto of Manchester who kindly read the manuscript and added many remarks that improved the original text. One of the authors (H.G.) is indebted to Professor Witold Brostow of the Center for Materials Characterization, University of North Texas, Denton for hospitality and A.C. Ramamurthy of the Ford Motor Co. for providing facilities that helped this paper to be written.

12
References and Notes

1. For the process of graph-theoretical evolution of trees, which is analogous to the aggregation described by Eq. (1), the Smoluchowski equation was rederived rigorously, see Balinska K, Galina H, Quintas LV, Szymanski J (1996) Discr Appl Maths 67:13
2. Pauling L, Pauling P (1975) Chemistry. Freeman, San Francisco 1975
3. Smoluchowski M v (1916) Phys Z 17: 585
4. Smoluchowski M v (1917) Z Chem Phys 92: 129
5. Drake R (1972) In: Hidy GM, Brook JR (eds) Topics in current aerosol research. Pergamon, Oxford
6. Kuchanov SJ (1978) Methods of kinetic calculations in polymer chemistry (in Russian). Izd Chimia, Moscow
7. Ziff RM.(1984) In: Family F, Landau DP (eds) Kinetics of aggregation and gelation. Elsevier Sci Publ BV
8. The 1/2 comes from the probability of collision of species of the same size which is proportional to,

$$\binom{Nc_k}{2} = \frac{1}{2} Nc_k (Nc_k - 1) \approx \frac{1}{2}(Nc_k)^2$$

while for i- and j mer ($i\,1j$) it is proportional to $\quad N^2 c_i c_j$

9. Flory JP (1953) Principles of polymer chemistry. Cornell Univ Press, Ithaca

10. An equivalent generating function defined as

$$H'(t,x) = \sum_k c_k e^{kx}$$

can also be used

11. One can arrive at equivalent result by differentiating $H(t,x)$ with respect to x followed by putting $x=0$, since

$$c_i = \frac{1}{i!}\left(\frac{\partial^i H}{\partial x^i}\right)\bigg|_{x=0}$$

12. Gordon M, Scantlebury GR (1966) Proc R Soc London A 292: 380
13. Gordon M, Scantlebury GR (1964) Trans Faraday Soc 6: 604
14. Gordon M (1962) Proc Roy Soc (London) A268: 240
15. Dušek K, Prins W (1969) Adv Polym Sci 6: 1
16. Galina H (1995) Makromol Theory Simul 4: 801
17. Cheng K Ch, Chiu WY (1993) Macromolecules 26: 4658, 4665
18. Flory PJ (1946) Chem Rev 39: 137
19. Stockmayer WH (1943) J Chem Phys 11: 45
20. The appearance of a giant component in the evolution of random graphs, which corresponds to gelation in polymer science, was discovered to their surprise by Erdös P, Renyi A. ([34] below) more than a decade after Flory
21. Spouge JL (1983) Macromolecules 16: 121
22. McLeod JB (1962) Quart J Math (Oxford) 13: 119, 193
23. Drake RL (1972) In: Hidy GM, Brock GR (eds) Topics in current aerosol research, vol 3, pt 2
24. Ziff RM, Ernst MH, Hendriks EM (1984) J Colloid Interface Sci 100: 220
25. Drake RL, Wright TJ (1972) J Atmos Sci 29: 548
26. Treat RP (1990) J Phys A: Math Gen 23: 3003
27. Spouge JL (1983) J Stat Phys 31: 363
28. Ziff RM, Stell G (1980) J Chem Phys 73: 3492
29. Van Dongen PGJ, Ernst MH (1987) J Colloid Interface Sci 115: 27
30. Van Dongen PGJ, Ernst MH (1986) J Stat Phys 44: 785
31. Leyvraz F (1984) In: Family F, Landau DP (eds) Kinetics aggregation and gelation. Elsevier Sci Publ BV
32. Leyvraz F, Tschudi HR (1981) J Phys A: Math Gen 14: 3389; (1982) J Phys A: Math Gen 15: 1951
33. Hendriks EM, Ernst MH, Ziff RM (1983) J Stat Phys 31: 519
34. Erdös P, Renýi A (1959) I Publ Math Debrecen 6: 290; (1960) MTA Mat Kut Int Kozl 5: 17
35. Buffet E, Pulé JV (1991) J Stat Phys 64: 87
36. Stockmayer WH (1943) J Chem Phys 11: 45
37. Whittle P (1965): Proc Camb Phil Soc 61: 475
38. Whittle P (1965) Proc R Soc London A 285: 501
39. Spouge JL (1983) J Phys A: Math Gen 16: 767
40. Stockmayer W H (1943) J Chem Phys 11: 45
41. Macosko CW, Miller DR (1976) Macromolecules 9: 199
42. Miller DR, Macosko CW (1976) Macromolecules 9: 206
43. Kuchanov SI, Povolotskaya Ye S (1982) Vysokomol Soed A24(10): 2179, 2190
44. Galina H, Szustalewicz A (1989) Macromolecules 22I 3124
45. Sarmoria C, Miller DR (1991) Macromolecules 24: 1833
46. Faliagas AC (1993) Macromolecules 26: 3838
47. Kuchanov SI (1979) Dokl Akad Nauk SSSR 249: 899
48. Galina H, Szustalewicz A (1990) Macromolecules 23: 3833
49. Galina H, Kaczmarski K, Para B, Sanecka B (1992) Makromol Chem Theory Simul 1: 37

50. Mikes J, Dušek K (1982) Macromolecules 15: 185
51. Šomvársky J, Dušek K (1994) Polym Bull (Berlin) 33: 369; 377
52. Galina H, Lechowicz J (1995) Comput Polym Sci
53. Dušek K (1979) Polym Bull (Berlin) 1: 523
54. Ziff RM (1980) J Stat Phys 23: 241
55. Good IJ (1963) Proc R Soc London A 272: 54
56. Van Dongen PGJ, Ernst MH, Ziff RM (1983) J Stat Phys 31: 519
57. Fukutomi T, Koshiro Y (1996) Polym J 28: 758
58. Irzhak TF, Peregudov NI, Irzhak VI, Rozenberg BA (1993) Vysokomol. Soed B 35: 905, 1545
59. Jackobson H, Stockmayer WH (1950) J Chem Phys 18: 1800
60. Flory P J (1969) Statistical mechanics of chain molecules. Interscience, New York
61. Friedman B, O'Shaughnessy B (1989) Phys Rev A 40: 5950
62. Lu B, Bak TA (1988) In: Kramer O (ed) Biological and synthetic polymer networks. Elsevier, Amsterdam
63. Galina H (1991) Makromol Chem Macromol Symp 40: 45
64. Galina H, Lechowicz J, Para B, Sanecka B (1994) Polimery (Warsaw) 39: 429
65. Müller H (1928) Kolloidchemische Beiheft 27: 223
66. Lushnikov AA (1974) J Colloid Interface Sci 48: 400; (1978) Izv Atm Ok Fiz 14: 738
67. Gabellini Y, Meunier J-L (1992) J Phys A: Math Gen 25: 3683
68. Kreer M, Penrose O (1994) J Stat Phys 75: 389
69. Van Dongen PGJ, Ernst MH (1987) J Stat Phys 49: 879
70. Stauffer D, Coniglio A, Adam M (1982) Adv Polym Sci 44: 103
71. Villarica M, Casey MJ, Goodisman J, Chaiken J (1992) J Chem Phys 98: 4610
72 Krivitsky DS (1996) J Phys A: Math Gen 28: 2025
73. Ziff RM, McGrady ED (1985) J Phys A: Math Gen 18: 3027
74. Ball JM, Carr J (1990) J Stat Phys 61: 203
75. Costas ME, Moreau M, Vicente L (1995) J Phys A: Math Gen 28: 2981
76. da Costa FP (1995) J Math Anal Appl 192:892

Editor: Prof. K. Dušek
Received: June 1997

Author Index Volumes 101–137

Author Index Volumes 1–100 see Volume 100

Adolf, D. B. see Ediger, M. D.: Vol. 116, pp. 73-110.
Aharoni, S. M. and *Edwards, S. F.*: Rigid Polymer Networks. Vol. 118, pp. 1-231.
Améduri, B., Boutevin, B. and *Gramain, P.*: Synthesis of Block Copolymers by Radical Polymerization and Telomerization. Vol. 127, pp. 87-142.
Améduri, B. and *Boutevin, B.*: Synthesis and Properties of Fluorinated Telechelic Monodispersed Compounds. Vol. 102, pp. 133-170.
Amselem, S. see Domb, A. J.: Vol. 107, pp. 93-142.
Andrady, A. L.: Wavelenght Sensitivity in Polymer Photodegradation. Vol. 128, pp. 47-94.
Andreis, M. and *Koenig, J. L.*: Application of Nitrogen-15 NMR to Polymers. Vol. 124, pp. 191-238.
Angiolini, L. see Carlini, C.: Vol. 123, pp. 127-214.
Anseth, K. S., Newman, S. M. and *Bowman, C. N.*: Polymeric Dental Composites: Properties and Reaction Behavior of Multimethacrylate Dental Restorations. Vol. 122, pp. 177-218.
Armitage, B. A. see O'Brien, D. F.: Vol. 126, pp. 53-58.
Arndt, M. see Kaminski, W.: Vol. 127, pp. 143-187.
Arnold Jr., F. E. and *Arnold, F. E.*: Rigid-Rod Polymers and Molecular Composites. Vol. 117, pp. 257-296.
Arshady, R.: Polymer Synthesis via Activated Esters: A New Dimension of Creativity in Macromolecular Chemistry. Vol. 111, pp. 1-42.

Bahar, I., Erman, B. and *Monnerie, L.*: Effect of Molecular Structure on Local Chain Dynamics: Analytical Approaches and Computational Methods. Vol. 116, pp. 145-206.
Baltá-Calleja, F. J., González Arche, A., Ezquerra, T. A., Santa Cruz, C., Batallón, F., Frick, B. and *López Cabarcos, E.*: Structure and Properties of Ferroelectric Copolymers of Poly(vinylidene) Fluoride. Vol. 108, pp. 1-48.
Barshtein, G. R. and *Sabsai, O. Y.*: Compositions with Mineralorganic Fillers. Vol. 101, pp. 1-28.
Batallán, F. see Baltá-Calleja, F. J.: Vol. 108, pp. 1-48.
Barton, J. see Hunkeler, D.: Vol. 112, pp. 115-134.
Bell, C. L. and *Peppas, N. A.*: Biomedical Membranes from Hydrogels and Interpolymer Complexes. Vol. 122, pp. 125-176.
Bellon-Maurel, A. see Calmon-Decriaud, A.: Vol. 135, pp. 207-226.
Bennett, D. E. see O'Brien, D. F.: Vol. 126, pp. 53-84.
Berry, G.C.: Static and Dynamic Light Scattering on Moderately Concentraded Solutions: Isotropic Solutions of Flexible and Rodlike Chains and Nematic Solutions of Rodlike Chains. Vol. 114, pp. 233-290.
Bershtein, V. A. and *Ryzhov, V. A.*: Far Infrared Spectroscopy of Polymers. Vol. 114, pp. 43-122.
Bigg, D. M.: Thermal Conductivity of Heterophase Polymer Compositions. Vol. 119, pp. 1-30.
Binder, K.: Phase Transitions in Polymer Blends and Block Copolymer Melts: Some Recent Developments. Vol. 112, pp. 115-134.
Bird, R. B. see Curtiss, C. F.: Vol. 125, pp. 1-102.

Biswas, M. and *Mukherjee, A.*: Synthesis and Evaluation of Metal-Containing Polymers. Vol. 115, pp. 89-124.
Boutevin, B. and *Robin, J. J.*: Synthesis and Properties of Fluorinated Diols. Vol. 102. pp. 105-132.
Boutevin, B. see Amédouri, B.: Vol. 102, pp. 133-170.
Boutevin, B. see Améduri, B.: Vol. 127, pp. 87-142.
Bowman, C. N. see Anseth, K. S.: Vol. 122, pp. 177-218.
Boyd, R. H.: Prediction of Polymer Crystal Structures and Properties. Vol. 116, pp. 1-26.
Bronnikov, S. V., Vettegren, V. I. and *Frenkel, S. Y.*: Kinetics of Deformation and Relaxation in Highly Oriented Polymers. Vol. 125, pp. 103-146.
Bruza, K. J. see Kirchhoff, R. A.: Vol. 117, pp. 1-66.
Burban, J. H. see Cussler, E. L.: Vol. 110, pp. 67-80.

Calmon-Decriaud, A. Bellon-Maurel, V., Silvestre, F.: Standard Methods for Testing the Aerobic Biodegradation of Polymeric Matcrials. Vol 135, pp. 207-226.
Cameron, N. R. and *Sherrington, D. C.*: High Internal Phase Emulsions (HIPEs)-Structure, Properties and Use in Polymer Preparation. Vol. 126, pp. 163-214.
Candau, F. see Hunkeler, D.: Vol. 112, pp. 115-134.
Canelas, D. A. and *DeSimone, J. M.*: Polymerizations in Liquid and Supercritical Carbon Dioxide. Vol. 133, pp. 103-140.
Capek, I.: Kinetics of the Free-Radical Emulsion Polymerization of Vinyl Chloride. Vol. 120, pp. 135-206.
Carlini, C. and *Angiolini, L.*: Polymers as Free Radical Photoinitiators. Vol. 123, pp. 127-214.
Casas-Vazquez, J. see Jou, D.: Vol. 120, pp. 207-266.
Chandrasekhar, V.: Polymer Solid Electrolytes: Synthesis and Structure. Vol 135, pp. 139-206
Chen, P. see Jaffe, M.: Vol. 117, pp. 297-328.
Choe, E.-W. see Jaffe, M.: Vol. 117, pp. 297-328.
Chow, T. S.: Glassy State Relaxation and Deformation in Polymers. Vol. 103, pp. 149-190.
Chung, T.-S. see Jaffe, M.: Vol. 117, pp. 297-328.
Connell, J. W. see Hergenrother, P. M.: Vol. 117, pp. 67-110.
Criado-Sancho, M. see Jou, D.: Vol. 120, pp. 207-266.
Curro, J.G. see Schweizer, K.S.: Vol. 116, pp. 319-378.
Curtiss, C. F. and *Bird, R. B.*: Statistical Mechanics of Transport Phenomena: Polymeric Liquid Mixtures. Vol. 125, pp. 1-102.
Cussler, E. L., Wang, K. L. and *Burban, J. H.*: Hydrogels as Separation Agents. Vol. 110, pp. 67-80.

DeSimone, J. M. see Canelas D. A.: Vol. 133, pp. 103-140.
DiMari, S. see Prokop, A.: Vol. 136, pp. 1-52.
Dimonie, M. V. see Hunkeler, D.: Vol. 112, pp. 115-134.
Dodd, L. R. and *Theodorou, D. N.*: Atomistic Monte Carlo Simulation and Continuum Mean Field Theory of the Structure and Equation of State Properties of Alkane and Polymer Melts. Vol. 116, pp. 249-282.
Doelker, E.: Cellulose Derivatives. Vol. 107, pp. 199-266.
Domb, A. J., Amselem, S., Shah, J. and *Maniar, M.*: Polyanhydrides: Synthesis and Characterization. Vol.107, pp. 93-142.
Dubrovskii, S. A. see Kazanskii, K. S.: Vol. 104, pp. 97-134.
Dunkin, I. R. see Steinke, J.: Vol. 123, pp. 81-126.

Economy, J. and *Goranov, K.*: Thermotropic Liquid Crystalline Polymers for High Performance Applications. Vol. 117, pp. 221-256.
Ediger, M. D. and *Adolf, D. B.*: Brownian Dynamics Simulations of Local Polymer Dynamics. Vol. 116, pp. 73-110.
Edwards, S. F. see Aharoni, S. M.: Vol. 118, pp. 1-231.
Endo, T. see Yagci, Y.: Vol. 127, pp. 59-86.

Erman, B. see Bahar, I.: Vol. 116, pp. 145-206.
Ewen, B, Richter, D.: Neutron Spin Echo Investigations on the Segmental Dynamics of Polymers in Melts, Networks and Solutions. Vol. 134, pp. 1-130.
Ezquerra, T. A. see Baltá-Calleja, F. J.: Vol. 108, pp. 1-48.

Fendler, J.H.: Membrane-Mimetic Approach to Advanced Materials. Vol. 113, pp. 1-209.
Fetters, L. J. see Xu, Z.: Vol. 120, pp. 1-50.
Förster, S. and *Schmidt, M.*: Polyelectrolytes in Solution. Vol. 120, pp. 51-134.
Frenkel, S. Y. see Bronnikov, S. V.: Vol. 125, pp. 103-146.
Frick, B. see Baltá-Calleja, F. J.: Vol. 108, pp. 1-48.
Fridman, M. L.: see Terent'eva, J. P.: Vol. 101, pp. 29-64.
Funke, W.: Microgels-Intramolecularly Crosslinked Macromolecules with a Globular Structure. Vol. 136, pp. 137-232.

Galina, H.: Mean-Field Kinetic Modeling of Polymerization: The Smoluchowski Coagulation Equation. Vol. 137, pp. 135-172.
Ganesh, K. see Kishore, K.: Vol. 121, pp. 81-122.
Geckeler, K. E. see Rivas, B.: Vol. 102, pp. 171-188.
Geckeler, K. E.: Soluble Polymer Supports for Liquid-Phase Synthesis. Vol. 121, pp. 31-80.
Gehrke, S. H.: Synthesis, Equilibrium Swelling, Kinetics Permeability and Applications of Environmentally Responsive Gels. Vol. 110, pp. 81-144.
Godovsky, D. Y.: Electron Behavior and Magnetic Properties Polymer-Nanocomposites. Vol. 119, pp. 79-122.
González Arche, A. see Baltá-Calleja, F. J.: Vol. 108, pp. 1-48.
Goranov, K. see Economy, J.: Vol. 117, pp. 221-256.
Gramain, P. see Améduri, B.: Vol. 127, pp. 87-142.
Grosberg, A. and *Nechaev, S.*: Polymer Topology. Vol. 106, pp. 1-30.
Grubbs, R., Risse, W. and *Novac, B.*: The Development of Well-defined Catalysts for Ring-Opening Olefin Metathesis. Vol. 102, pp. 47-72.
van Gunsteren, W. F. see Gusev, A. A.: Vol. 116, pp. 207-248.
Gusev, A. A., Müller-Plathe, F., van Gunsteren, W. F. and *Suter, U. W.*: Dynamics of Small Molecules in Bulk Polymers. Vol. 116, pp. 207-248.
Guillot, J. see Hunkeler, D.: Vol. 112, pp. 115-134.
Guyot, A. and *Tauer, K.*: Reactive Surfactants in Emulsion Polymerization. Vol. 111, pp. 43-66.

Hadjichristidis, N. see Xu, Z.: Vol. 120, pp. 1-50.
Hadjichristidis, N. see Pitsikalis, M.: Vol. 135, pp. 1-138.
Hall, H. K. see Penelle, J.: Vol. 102, pp. 73-104.
Hammouda, B.: SANS from Homogeneous Polymer Mixtures: A Unified Overview. Vol. 106, pp. 87-134.
Harada, A.: Design and Construction of Supramolecular Architectures Consisting of Cyclodextrins and Polymers. Vol. 133, pp. 141-192.
Haralson, M. A. see Prokop, A.: Vol. 136, pp. 1-52.
Hedrick, J. L. see Hergenrother, P. M.: Vol. 117, pp. 67-110.
Heller, J.: Poly (Ortho Esters). Vol. 107, pp. 41-92.
Hemielec, A. A. see Hunkeler, D.: Vol. 112, pp. 115-134.
Hergenrother, P. M., Connell, J. W., Labadie, J. W. and *Hedrick, J. L.*: Poly(arylene ether)s Containing Heterocyclic Units. Vol. 117, pp. 67-110.
Hiramatsu, N. see Matsushige, M.: Vol. 125, pp. 147-186.
Hirasa, O. see Suzuki, M.: Vol. 110, pp. 241-262.
Hirotsu, S.: Coexistence of Phases and the Nature of First-Order Transition in Poly-N-isopropylacrylamide Gels. Vol. 110, pp. 1-26.
Hunkeler, D., Candau, F., Pichot, C., Hemielec, A. E., Xie, T. Y., Barton, J., Vaskova, V., Guillot, J., Dimonie, M. V., Reichert, K. H.: Heterophase Polymerization: A Physical and Kinetic Comparision and Categorization. Vol. 112, pp. 115-134.

Hunkeler, D. see Prokop, A.: Vol. 136, pp. 1-52; 53-74.

Ichikawa, T. see Yoshida, H.: Vol. 105, pp. 3-36.
Ihara, E. see Yasuda, H.: Vol. 133, pp. 53-102.
Ikada, Y. see Uyama, Y.: Vol. 137, pp. 1-40.
Ilavsky, M.: Effect on Phase Transition on Swelling and Mechanical Behavior of Synthetic Hydrogels. Vol. 109, pp. 173-206.
Inomata, H. see Saito, S.: Vol. 106, pp. 207-232.
Irie, M.: Stimuli-Responsive Poly(N-isopropylacrylamide), Photo- and Chemical-Induced Phase Transitions. Vol. 110, pp. 49-66.
Ise, N. see Matsuoka, H.: Vol. 114, pp. 187-232.
Ivanov, A. E. see Zubov, V. P.: Vol. 104, pp. 135-176.

Jaffe, M., Chen, P., Choe, E.-W., Chung, T.-S. and *Makhija, S.*: High Performance Polymer Blends. Vol. 117, pp. 297-328.
Joos-Müller, B. see Funke, W.: Vol. 136, pp. 137-232.
Jou, D., Casas-Vazquez, J. and *Criado-Sancho, M.*: Thermodynamics of Polymer Solutions under Flow: Phase Separation and Polymer Degradation. Vol. 120, pp. 207-266.

Kaetsu, I.: Radiation Synthesis of Polymeric Materials for Biomedical and Biochemical Applications. Vol. 105, pp. 81-98.
Kaminski, W. and *Arndt, M.*: Metallocenes for Polymer Catalysis. Vol. 127, pp. 143-187.
Kammer, H. W., Kressler, H. and *Kummerloewe, C.*: Phase Behavior of Polymer Blends - Effects of Thermodynamics and Rheology. Vol. 106, pp. 31-86.
Kandyrin, L. B. and *Kuleznev, V. N.*: The Dependence of Viscosity on the Composition of Concentrated Dispersions and the Free Volume Concept of Disperse Systems. Vol. 103, pp. 103-148.
Kaneko, M. see Ramaraj, R.: Vol. 123, pp. 215-242.
Kang, E. T., Neoh, K. G. and *Tan, K. L.*: X-Ray Photoelectron Spectroscopic Studies of Electroactive Polymers. Vol. 106, pp. 135-190.
Kato, K. see Uyama, Y.: Vol. 137, pp. 1-40.
Kazanskii, K. S. and *Dubrovskii, S. A.*: Chemistry and Physics of „Agricultural" Hydrogels. Vol. 104, pp. 97-134.
Kennedy, J. P. see Majoros, I.: Vol. 112, pp. 1-113.
Khokhlov, A., Starodybtzev, S. and *Vasilevskaya, V.*: Conformational Transitions of Polymer Gels: Theory and Experiment. Vol. 109, pp. 121-172.
Kilian, H. G. and *Pieper, T.*: Packing of Chain Segments. A Method for Describing X-Ray Patterns of Crystalline, Liquid Crystalline and Non-Crystalline Polymers. Vol. 108, pp. 49-90.
Kishore, K. and *Ganesh, K.*: Polymers Containing Disulfide, Tetrasulfide, Diselenide and Ditelluride Linkages in the Main Chain. Vol. 121, pp. 81-122.
Kitamaru, R.: Phase Structure of Polyethylene and Other Crystalline Polymers by Solid-State ^{13}C/MNR. Vol. 137, pp 41-102.
Klier, J. see Scranton, A. B.: Vol. 122, pp. 1-54.
Kobayashi, S., Shoda, S. and *Uyama, H.*: Enzymatic Polymerization and Oligomerization. Vol. 121, pp. 1-30.
Koenig, J. L. see Andreis, M.: Vol. 124, pp. 191-238.
Kokufuta, E.: Novel Applications for Stimulus-Sensitive Polymer Gels in the Preparation of Functional Immobilized Biocatalysts. Vol. 110, pp. 157-178.
Konno, M. see Saito, S.: Vol. 109, pp. 207-232.
Kopecek, J. see Putnam, D.: Vol. 122, pp. 55-124.
Koßmehl, G. see Schopf, G.: Vol. 129, pp. 1-145.
Kressler, J. see Kammer, H. W.: Vol. 106, pp. 31-86.
Kirchhoff, R. A. and *Bruza, K. J.*: Polymers from Benzocyclobutenes. Vol. 117, pp. 1-66.
Kuchanov, S. I.: Modern Aspects of Quantitative Theory of Free-Radical Copolymerization. Vol. 103, pp. 1-102.

Kuleznev, V. N. see *Kandyrin, L. B.*: Vol. 103, pp. 103-148.
Kulichkhin, S. G. see *Malkin, A. Y.*: Vol. 101, pp. 217-258.
Kummerloewe, C. see *Kammer, H. W.*: Vol. 106, pp. 31-86.
Kuznetsova, N. P. see *Samsonov, G. V.*: Vol. 104, pp. 1-50. Labadie, J. W. see *Hergenrother, P. M.*: Vol. 117, pp. 67-110.

Lamparski, H. G. see *O´Brien, D. F.*: Vol. 126, pp. 53-84.
Laschewsky, A.: Molecular Concepts, Self-Organisation and Properties of Polysoaps. Vol. 124, pp. 1-86.
Laso, M. see *Leontidis, E.*: Vol. 116, pp. 283-318.
Lazár, M. and *RychlΩ, R.*: Oxidation of Hydrocarbon Polymers. Vol. 102, pp. 189-222.
Lechowicz, J. see *Galina, H.*: Vol. 137, pp. 135-172.
Lenz, R. W.: Biodegradable Polymers. Vol. 107, pp. 1-40.
Leontidis, E., de Pablo, J. J., Laso, M. and *Suter, U. W.*: A Critical Evaluation of Novel Algorithms for the Off-Lattice Monte Carlo Simulation of Condensed Polymer Phases. Vol. 116, pp. 283-318.
Lesec, J. see *Viovy, J.-L.*: Vol. 114, pp. 1-42.
Liang, G. L. see *Sumpter, B. G.*: Vol. 116, pp. 27-72.
Lin, J. and *Sherrington, D. C.*: Recent Developments in the Synthesis, Thermostability and Liquid Crystal Properties of Aromatic Polyamides. Vol. 111, pp. 177-220.
López Cabarcos, E. see *Baltá-Calleja, F. J.*: Vol. 108, pp. 1-48.

Majoros, I., Nagy, A. and *Kennedy, J. P.*: Conventional and Living Carbocationic Polymerizations United. I. A Comprehensive Model and New Diagnostic Method to Probe the Mechanism of Homopolymerizations. Vol. 112, pp. 1-113.
Makhija, S. see *Jaffe, M.*: Vol. 117, pp. 297-328.
Malkin, A. Y. and *Kulichkhin, S. G.*: Rheokinetics of Curing. Vol. 101, pp. 217-258.
Maniar, M. see *Domb, A. J.*: Vol. 107, pp. 93-142.
Mashima, K., Nakayama, Y. and *Nakamura, A.*: Recent Trends in Polymerization of a-Olefins Catalyzed by Organometallic Complexes of Early Transition Metals. Vol. 133, pp. 1-52.
Matsumoto, A.: Free-Radical Crosslinking Polymerization and Copolymerization of Multivinyl Compounds. Vol. 123, pp. 41-80.
Matsumoto, A. see *Otsu, T.*: Vol. 136, pp. 75-138.
Matsuoka, H. and *Ise, N.*: Small-Angle and Ultra-Small Angle Scattering Study of the Ordered Structure in Polyelectrolyte Solutions and Colloidal Dispersions. Vol. 114, pp. 187-232.
Matsushige, K., Hiramatsu, N. and *Okabe, H.*: Ultrasonic Spectroscopy for Polymeric Materials. Vol. 125, pp. 147-186.
Mattice, W. L. see *Rehahn, M.*: Vol. 131/132, pp. 1-475.
Mays, W. see *Xu, Z.*: Vol. 120, pp. 1-50.
Mays, J. W. see *Pitsikalis, M.*: Vol. 135, pp. 1-138.
Mikos, A. G. see *Thomson, R. C.*: Vol. 122, pp. 245-274.
Miyasaka, K.: PVA-Iodine Complexes: Formation, Structure and Properties. Vol. 108. pp. 91-130.
Monnerie, L. see *Bahar, I.*: Vol. 116, pp. 145-206.
Morishima, Y.: Photoinduced Electron Transfer in Amphiphilic Polyelectrolyte Systems. Vol. 104, pp. 51-96.
Mours, M. see *Winter, H. H.*: Vol. 134, pp. 165-234.
Müllen, K. see *Scherf, U.*: Vol. 123, pp. 1-40.
Müller-Plathe, F. see *Gusev, A. A.*: Vol. 116, pp. 207-248.
Mukerherjee, A. see *Biswas, M.*: Vol. 115, pp. 89-124.
Mylnikov, V.: Photoconducting Polymers. Vol. 115, pp. 1-88.

Nagy, A. see *Majoros, I.*: Vol. 112, pp. 1-11.
Nakamura, A. see *Mashima, K.*: Vol. 133, pp. 1-52.
Nakayama, Y. see *Mashima, K.*: Vol. 133, pp. 1-52.
Narasinham, B., Peppas, N. A.: The Physics of Polymer Dissolution: Modeling Approaches and Experimental Behavior. Vol. 128, pp. 157-208.
Nechaev, S. see *Grosberg, A.*: Vol. 106, pp. 1-30.

Neoh, K. G. see *Kang, E. T.*: Vol. 106, pp. 135-190.
Newman, S. M. see *Anseth, K. S.*: Vol. 122, pp. 177-218.
Nijenhuis, K. te: Thermoreversible Networks. Vol. 130, pp. 1-252.
Noid, D. W. see *Sumpter, B. G.*: Vol. 116, pp. 27-72.
Novac, B. see *Grubbs, R.*: Vol. 102, pp. 47-72.
Novikov, V. V. see *Privalko, V. P.*: Vol. 119, pp. 31-78.

O'Brien, D. F., Armitage, B. A., Bennett, D. E. and *Lamparski, H. G.*: Polymerization and Domain Formation in Lipid Assemblies. Vol. 126, pp. 53-84.
Ogasawara, M.: Application of Pulse Radiolysis to the Study of Polymers and Polymerizations. Vol. 105, pp. 37-80.
Okabe, H. see *Matsushige, K.*: Vol. 125, pp. 147-186.
Okada, M.: Ring-Opening Polymerization of Bicyclic and Spiro Compounds. Reactivities and Polymerization Mechanisms. Vol. 102, pp. 1-46.
Okano, T.: Molecular Design of Temperature-Responsive Polymers as Intelligent Materials. Vol. 110, pp. 179-198.
Okay, O. see *Funke, W.*: Vol. 136, pp. 137-232.
Onuki, A.: Theory of Phase Transition in Polymer Gels. Vol. 109, pp. 63-120.
Osad'ko, I.S.: Selective Spectroscopy of Chromophore Doped Polymers and Glasses. Vol. 114, pp. 123-186.
Otsu, T., Matsumoto, A.: Controlled Synthesis of Polymers Using the Iniferter Technique: Developments in Living Radical Polymerization. Vol. 136, pp. 75-138.

de Pablo, J. J. see *Leontidis, E.*: Vol. 116, pp. 283-318.
Padias, A. B. see *Penelle, J.*: Vol. 102, pp. 73-104.
Pascault, J.-P. see *Williams, R. J. J.*: Vol. 128, pp. 95-156.
Pasch, H.: Analysis of Complex Polymers by Interaction Chromatography. Vol. 128, pp. 1-46.
Penelle, J., Hall, H. K., Padias, A. B. and *Tanaka, H.*: Captodative Olefins in Polymer Chemistry. Vol. 102, pp. 73-104.
Peppas, N. A. see *Bell, C. L.*: Vol. 122, pp. 125-176.
Peppas, N. A. see *Narasimhan, B.*: Vol. 128, pp. 157-208.
Pichot, C. see *Hunkeler, D.*: Vol. 112, pp. 115-134.
Pieper, T. see *Kilian, H. G.*: Vol. 108, pp. 49-90.
Pispas, S. see *Pitsikalis, M.*: Vol. 135, pp. 1-138.
Pitsikalis, M., Pispas, S., Mays, J. W., Hadjichristidis, N.: Nonlinear Block Copolymer Architectures. Vol. 135, pp. 1-138.
Pospíšil, J.: Functionalized Oligomers and Polymers as Stabilizers for Conventional Polymers. Vol. 101, pp. 65-168.
Pospíšil, J.: Aromatic and Heterocyclic Amines in Polymer Stabilization. Vol. 124, pp. 87-190.
Powers, A. C. see *Prokop, A.*: Vol. 136, pp. 53-74.
Priddy, D. B.: Recent Advances in Styrene Polymerization. Vol. 111, pp. 67-114.
Priddy, D. B.: Thermal Discoloration Chemistry of Styrene-co-Acrylonitrile. Vol. 121, pp. 123-154.
Privalko, V. P. and *Novikov, V. V.*: Model Treatments of the Heat Conductivity of Heterogeneous Polymers. Vol. 119, pp 31-78.
Prokop, A., Hunkeler, D., Powers, A. C., Whitesell, R. R., Wang, T. G.: Water Soluble Polymers for Immunoisolation II: Evaluation of Multicomponent Microencapsulation Systems. Vol. 136, pp. 53-74.
Prokop, A., Hunkeler, D., DiMari, S., Haralson, M. A., Wang, T. G.: Water Soluble Polymers for Immunoisolation I: Complex Coacervation and Cytotoxicity. Vol. 136, pp. 1-52.
Putnam, D. and *Kopecek, J.*: Polymer Conjugates with Anticancer Acitivity. Vol. 122, pp. 55-124.

Ramaraj, R. and *Kaneko, M.*: Metal Complex in Polymer Membrane as a Model for Photosynthetic Oxygen Evolving Center. Vol. 123, pp. 215-242.
Rangarajan, B. see *Scranton, A. B.*: Vol. 122, pp. 1-54.
Reichert, K. H. see *Hunkeler, D.*: Vol. 112, pp. 115-134.

Rehahn, M., Mattice, W. L., Suter, U. W.: Rotational Isomeric State Models in Macromolecular Systems. Vol. 131/132, pp. 1-475.
Richter, D. see *Ewen, B.*: Vol. 134, pp.1-130.
Risse, W. see *Grubbs, R.*: Vol. 102, pp. 47-72.
Rivas, B. L. and *Geckeler, K. E.*: Synthesis and Metal Complexation of Poly(ethyleneimine) and Derivatives. Vol. 102, pp. 171-188.
Robin, J. J. see *Boutevin, B.*: Vol. 102, pp. 105-132.
Roe, R.-J.: MD Simulation Study of Glass Transition and Short Time Dynamics in Polymer Liquids. Vol. 116, pp. 111-114.
Rozenberg, B. A. see *Williams, R. J. J.*: Vol. 128, pp. 95-156.
Ruckenstein, E.: Concentrated Emulsion Polymerization. Vol. 127, pp. 1-58.
Rusanov, A. L.: Novel Bis (Naphtalic Anhydrides) and Their Polyheteroarylenes with Improved Processability. Vol. 111, pp. 115-176.
Rychlý, J. see *Lazár, M.*: Vol. 102, pp. 189-222.
Ryzhov, V. A. see *Bershtein, V. A.*: Vol. 114, pp. 43-122.

Sabsai, O. Y. see *Barshtein, G. R.*: Vol. 101, pp. 1-28.
Saburov, V. V. see *Zubov, V. P.*: Vol. 104, pp. 135-176.
Saito, S., Konno, M. and *Inomata, H.*: Volume Phase Transition of N-Alkylacrylamide Gels. Vol. 109, pp. 207-232.
Samsonov, G. V. and *Kuznetsova, N. P.*: Crosslinked Polyelectrolytes in Biology. Vol. 104, pp. 1-50.
Santa Cruz, C. see *Baltá-Calleja, F. J.*: Vol. 108, pp. 1-48.
Sato, T. and *Teramoto, A.*: Concentrated Solutions of Liquid-Christalline Polymers. Vol. 126, pp. 85-162.
Scherf, U. and *Müllen, K.*: The Synthesis of Ladder Polymers. Vol. 123, pp. 1-40.
Schmidt, M. see *Förster, S.*: Vol. 120, pp. 51-134.
Schopf, G. and *Koßmehl, G.*: Polythiophenes - Electrically Conductive Polymers. Vol. 129, pp. 1-145.
Schweizer, K. S.: Prism Theory of the Structure, Thermodynamics, and Phase Transitions of Polymer Liquids and Alloys. Vol. 116, pp. 319-378.
Scranton, A. B., Rangarajan, B. and *Klier, J.*: Biomedical Applications of Polyelectrolytes. Vol. 122, pp. 1-54.
Sefton, M. V. and *Stevenson, W. T. K.*: Microencapsulation of Live Animal Cells Using Polycrylates. Vol.107, pp. 143-198.
Shamanin, V. V.: Bases of the Axiomatic Theory of Addition Polymerization. Vol. 112, pp. 135-180.
Sherrington, D. C. see *Cameron, N. R.*, Vol. 126, pp. 163-214.
Sherrington, D. C. see *Lin, J.*: Vol. 111, pp. 177-220.
Sherrington, D. C. see *Steinke, J.*: Vol. 123, pp. 81-126.
Shibayama, M. see *Tanaka, T.*: Vol. 109, pp. 1-62.
Shiga, T.: Deformation and Viscoelastic Behavior of Polymer Gels in Electric Fields. Vol. 134, pp. 131-164.
Shoda, S. see *Kobayashi, S.*: Vol. 121, pp. 1-30.
Siegel, R. A.: Hydrophobic Weak Polyelectrolyte Gels: Studies of Swelling Equilibria and Kinetics. Vol. 109, pp. 233-268.
Silvestre, F. see *Calmon-Decriaud, A.*: Vol. 207, pp. 207-226.
Singh, R. P. see *Sivaram, S.*: Vol. 101, pp. 169-216.
Sivaram, S. and *Singh, R. P.*: Degradation and Stabilization of Ethylene-Propylene Copolymers and Their Blends: A Critical Review. Vol. 101, pp. 169-216.
Starodybtzev, S. see *Khokhlov, A.*: Vol. 109, pp. 121-172.
Steinke, J., Sherrington, D. C. and *Dunkin, I. R.*: Imprinting of Synthetic Polymers Using Molecular Templates. Vol. 123, pp. 81-126.
Stenzenberger, H. D.: Addition Polyimides. Vol. 117, pp. 165-220.
Stevenson, W. T. K. see *Sefton, M. V.*: Vol. 107, pp. 143-198.
Sumpter, B. G., Noid, D. W., Liang, G. L. and *Wunderlich, B.*: Atomistic Dynamics of Macromolecular Crystals. Vol. 116, pp. 27-72.

Suter, U. W. see Gusev, A. A.: Vol. 116, pp. 207-248.
Suter, U. W. see Leontidis, E.: Vol. 116, pp. 283-318.
Suter, U. W. see Rehahn, M.: Vol. 131/132, pp. 1-475.
Suzuki, A.: Phase Transition in Gels of Sub-Millimeter Size Induced by Interaction with Stimuli. Vol. 110, pp. 199-240.
Suzuki, A. and *Hirasa, O.*: An Approach to Artifical Muscle by Polymer Gels due to Micro-Phase Separation. Vol. 110, pp. 241-262.

Tagawa, S.: Radiation Effects on Ion Beams on Polymers. Vol. 105, pp. 99-116.
Tan, K. L. see Kang, E. T.: Vol. 106, pp. 135-190.
Tanaka, T. see Penelle, J.: Vol. 102, pp. 73-104.
Tanaka, H. and *Shibayama, M.*: Phase Transition and Related Phenomena of Polymer Gels. Vol. 109, pp. 1-62.
Tauer, K. see Guyot, A.: Vol. 111, pp. 43-66.
Teramoto, A. see Sato, T.: Vol. 126, pp. 85-162.
Terent´eva, J. P. and *Fridman, M. L.*: Compositions Based on Aminoresins. Vol. 101, pp. 29-64.
Theodorou, D. N. see Dodd, L. R.: Vol. 116, pp. 249-282.
Thomson, R. C., Wake, M. C., Yaszemski, M. J. and *Mikos, A. G.*: Biodegradable Polymer Scaffolds to Regenerate Organs. Vol. 122, pp. 245-274.
Tokita, M.: Friction Between Polymer Networks of Gels and Solvent. Vol. 110, pp. 27-48.
Tsuruta, T.: Contemporary Topics in Polymeric Materials for Biomedical Applications. Vol. 126, pp. 1-52.

Uyama, H. see Kobayashi, S.: Vol. 121, pp. 1-30.
Uyama, Y: Surface Modification of Polymers by Grafting. Vol. 137, pp. 1-40.

Vasilevskaya, V. see Khokhlov, A.: Vol. 109, pp. 121-172.
Vaskova, V. see Hunkeler, D.: Vol.:112, pp. 115-134.
Verdugo, P.: Polymer Gel Phase Transition in Condensation-Decondensation of Secretory Products. Vol. 110, pp. 145-156.
Vettegren, V. I.: see Bronnikov, S. V.: Vol. 125, pp. 103-146.
Viovy, J.-L. and *Lesec, J.*: Separation of Macromolecules in Gels: Permeation Chromatography and Electrophoresis. Vol. 114, pp. 1-42.
Volksen, W.: Condensation Polyimides: Synthesis, Solution Behavior, and Imidization Characteristics. Vol. 117, pp. 111-164.

Wake, M. C. see Thomson, R. C.: Vol. 122, pp. 245-274.
Wang, K. L. see Cussler, E. L.: Vol. 110, pp. 67-80.
Wang, T. G. see Prokop, A.: Vol. 136, pp.1-52; 53-74.
Whitesell, R. R. see Prokop, A.: Vol. 136, pp. 53-74.
Williams, R. J. J., Rozenberg, B. A., Pascault, J.-P.: Reaction Induced Phase Separation in Modified Thermosetting Polymers. Vol. 128, pp. 95-156.
Winter, H. H., Mours, M.: Rheology of Polymers Near Liquid-Solid Transitions. Vol. 134, pp. 165-234.
Wu, C.: Laser Light Scattering Characterization of Special Intractable Macromolecules in Solution. Vol 137, pp. 103-134.
Wunderlich, B. see Sumpter, B. G.: Vol. 116, pp. 27-72.

Xie, T. Y. see Hunkeler, D.: Vol. 112, pp. 115-134.
Xu, Z., Hadjichristidis, N., Fetters, L. J. and *Mays, J. W.*: Structure/Chain-Flexibility Relationships of Polymers. Vol. 120, pp. 1-50.

Yagci, Y. and *Endo, T.*: N-Benzyl and N-Alkoxy Pyridium Salts as Thermal and Photochemical Initiators for Cationic Polymerization. Vol. 127, pp. 59-86.
Yannas, I. V.: Tissue Regeneration Templates Based on Collagen-Glycosaminoglycan Copolymers. Vol. 122, pp. 219-244.
Yamaoka, H.: Polymer Materials for Fusion Reactors. Vol. 105, pp. 117-144.

Yasuda, H. and *Ihara, E.*: Rare Earth Metal-Initiated Living Polymerizations of Polar and Nonpolar Monomers. Vol. 133, pp. 53-102.
Yaszemski, M. J. see Thomson, R. C.: Vol. 122, pp. 245-274.
Yoshida, H. and *Ichikawa, T.*: Electron Spin Studies of Free Radicals in Irradiated Polymers. Vol. 105, pp. 3-36.

Zubov, V. P., Ivanov, A. E. and *Saburov, V. V.*: Polymer-Coated Adsorbents for the Separation of Biopolymers and Particles. Vol. 104, pp. 135-176.

Subject Index

Absolute method 107
Absorption bands 93
Activation energy 144, 146
Acyclic molecules (aggregates) 139, 155, 157, 165
Adhesion energy 18
Aggregate size 137, 138, 155, 162, 168, 169
Aggregation physics 156, 165
Aggregation processes 137, 138, 154, 155, 168
Aggregation rate 154
Aggregation-fragmentation model 170
Alternating copolymerization 145, 146, 150
Amorphous region 43
Analogue-to-digital 115
Angular dependence 108, 118
Anticoagulant 28
Apparent weight distribution function 125
Arterio-venous shunt 24
Assemblage of structures 137
Atmospheric pressure 65, 66, 68
Autoantibody 29
Average degree of polymerization 139-150, 157, 161, 169
Average diffusion coefficient 110
Average molar mass 106, 118, 127
Average molecular weight 136
Avogadro constant 108, 129

Backward reaction 155
Bilinear kernel 151, 155, 165
Boltzmann constant 129
Bound solvent 97
Branched structures 152
Broad component 49
Broad-line ^{1}H NMR 42

$1B_2$ Carbon 77
Cascade theory 145, 146, 164
Chain composition distribution 125
Chain scission 5
Chain transfer agent (reagent) 8, 28
Chains 124-129

Chemical shift 43, 44
Chemical shift Hamiltonian 43, 44
Chemical shift spectrum 44
Chemical shift tensor 51
Chitosan 110, 121
Chromatography 103, 105, 107, 122, 129
Classical model 141
Clotting cascade 28
Clusters 127
Coagulation kernel 138, 140, 151-154, 162, 165-169
Coefficient of friction 31
Collision frequency function 168
Colloidal particle 106, 129
Combinatorial identity 155
Composite material 19
Concentration dependence 117
Connectivity 137
CONTIN 104, 110
Conversion degree 137, 142, 143, 156, 161
Copolymer 105, 108, 111, 117, 124, 125
Correlation function 46, 47
Critical time 162, 169
Cross-polarization, CP 44
Crosslinking 9
Crosslinking reactions 162
Crystalline component 42
Crystalline fraction 49
Crystalline lamella 49
Crystalline stem length, ξ_C 58
Cycle formation 160, 166
Cycles 157, 166, 167, 168
Cyclization reactions 165, 166, 167
Cyclo alkanes 63

DD 4
DD/MAS ^{13}C NMR 45
Debye screening length 18
Delay time 104, 109
Dendrimer 152
o-Dichlorobenzene 90, 91
Dielectric measurement 49

Differential equation 155, 158, 160, 161, 167
Differential scanning calorimetry, DSC 64, 72, 80
Differentiation 141, 145, 149, 150, 153, 158-164, 170
Diffusion 137
Diffusion second virial coefficient 104, 109
Dimensionality 154, 166
Dipolar interaction 46
Direct dipolar interaction, H_D 43, 44
Dissolution/filtration apparatus 112
DPPH method 7, 10
Dried gel 70, 74
DSC 64, 72, 80
Dummy variable 140, 158, 160, 161, 168
Dynamics LLS 103, 107, 109, 118, 122, 124-130

Elasticity 137
Electroactive polymer 13
Electroosmotic motion 16
Electrophoretic mobility 16
Elementary reactions 145
Ellipsometry 16
Elution time 105, 122, 123
End-group chemical analysis 107
Endothermal peak 73
Entrapment of platelet 27
Epoxy resins 162
Equilibrium distribution 155
Error analysis 119
Esterification 5
Ethyl branches 74
Ethyl groups 77, 78
Ethylene-butene copolymer 78
Evanescent-wave 17
Evolution time 96
Excess Rayleigh ratio 103, 108
Explicit distribution 142, 150, 154, 156, 162
Explicit solution 150, 154, 160, 161, 166
Exponent 154, 169, 160
Extended crystal 64
Extended molecular chain length 50
Extinction probability 164

Fiber period 93
FID 45
Field flow fractionation (FFF) 107
First-order electric field time correlation function 104, 109
Flory model 163, 164
Flory-Stockmayer model 151-155, 158-160
Flory's postulate 151, 156
Fluctuation 137, 138
Fluorocarbon polymers 104, 111

Foreign-body response 24
Fourier transform 46
Free induction decay, FID 45
Free solvent 97
Freely-jointed chain model 166

γ-Gauche effect 92-94
Gaussian 48
Gaussian chain model 166
Gel permeation chromatography (GPC) 107
Gel point 137, 151, 154, 160-166
Gelatin 124
Gelation 128, 137, 151, 155, 162-165, 169-171
Generalized Smoluchowski equation 169
Generating function 140, 145, 167, 170
Geometric factor 154
ggtgg conformation 63
ggtt sequence 63
Glassy state 42
Growth to infinity 163
Guiselin brush 15, 16
Gyration, radius of 104, 108

^1H dipolar decoupling, DD 44
Hagen-Poiseuille equation 17
Heat of fusion 80
Helical chain, 3_1- 86
Helical molecular chain conformation 84, 89, 92
Hemocompatible dialysis 28
Hexagonal crystal form 73
High resolution solid-state ^{13}C NMR 42
High-pressure crystallized polyethylene 24-30, 64-70
High-temperature LLS spectrometer 103, 111
Hindered rotational 49
Hizex 64
Hofmann degradation 28
Homogeneous kernel 168, 169
Homopolymerization 139, 142, 143, 151, 155, 167
Homopolymers 124, 125
Hydrodynamic size 122, 127
Hydrodynamic thickness 16
Hydrogenated polybutadiene 74-79

Immunoadsorbent 29
Indirect dipolar interaction, J_J 43, 44
Input parameters 136
Intensity-intensity time correlation 104, 109, 118
Interfacial component 42
Interfacial region 42
Interlamellar material 50

Subject Index

Intermediate component 42
Internuclear distance 47
Internuclear vector 44
Interspin vector 44
Intramolecular reactions 137, 139, 154, 164, 167
Intrinsic viscosity 105, 121
Inversion recovery pulse sequence 52
Ionizing radiation 5, 7
iPP crystal, α-form/β-form 84
IR spectrum 93
Isotactic polypropylene 84-89, 92
Isotacticity 85

Kapton 10
Kevlar 9, 20
Kinetic analysis 137, 138, 143-148, 155-163, 166
Kuhn length 18

Lagrange expansion 154, 155
Laplace inversion 104, 110, 118, 119
Laplace transform 165
Larmor frequency 47
Laser light scattering 103, 107, 108, 119, 122, 129, 131
Latex 129
Law of mass action 138
Least-squares fitting 56
Limited molecualar motion 59
Line shape 48
Line shape analysis 55
Line width 48, 104, 105, 109
Linear polyethylene, from dilute solution 61-64
– –, randomly distributed ethyl branches 74-79
Linear polymerization 139, 146, 166
Liquid-like amorphous component 49
Liquid-like mobility 49
Living polymerization 5
Local molecular movement 49
Longitudinal relaxation 46
Lorentzian 48
Lower critical solution temperature 17
Lozenge-shaped crystallites 61, 63
Lozenge-shaped single crystal 24
Lushnikov process 168

Magic angle spinning (MAS) 44
Magnetic transmittance 47
Magnetogyric ration 7, 47
Mark-Houwink equation 121
MAS 44
Mass distribution 129
Mass fraction 49

Mass fraction of interlamellar material 50
Matrix-assisted time-fly mass spectroscopy 107
Measured base line 109
Mechanical measurement 49
Mechanism of aggregation 154
Melting temperature 73
Membrane osmometry 107
Methine carbon 75
Method of characteristics 153, 154
Methyl carbon 77
Methylene carbon 75
Methylene carbon in ethyl branches, $1B_2$ carbon 77
Methylene chloride 80
Methylene group, Methylene sequence 49
Micro-Brownian motion 49
Mobility of solvent 96
Moleacular mobility 43
Molecualr dynamics 42
Molecular chain conformation 42
Molecular chain folding 50
Molecular dynamics 15
Molecular graph 154, 167, 170, 171
Molecular weight 136
– –, Z-average 107
Molecular weight dependence of phase structure 58
Molecular weight distribution 106, 107, 118-121, 124, 125
Moment of distribution 139-142, 145, 149-155, 160-164, 167
Monochromatic 107
Monoclinic crystal form 58
Monodisperse 103, 104, 109, 119
Monte Carlo calculations 162
Monte Carlo simulation 15
Most probable distribution 143
Mulitcomponent system 145, 147
Multiplicative kernel 165

Narrow component 49
Net scattering intensity 107
Neutron reflection 14, 15
Neutron scattering 14, 16, 42
Nitric acid degradation/g.p.c. technique 68
Noncrystalline amorphous component 42
Noncrystalline component 42
Noncrystalline interlamellar material 49
Normal vibrational mode 93
Normalization 139
Number-average chain length 69
Number-average crystal thickness, ξ_n 69
Numerical soutions 145, 148, 149, 161, 165, 168, 169

Orientation function 47
Orthorhombic crystalline form 51
Osmometry 80
–, vapour pressure 107
Osteoblast cell culture 35

Packing effects 84-86, 88, 94
n-Paraffins 51, 72
Partitioning of methyl carbon 77
Percutaneous implantation 33
Phase structure 42
Photo-initiator 12
Photon correlation 104, 109, 110
Photon correlation spectroscopy (PCS) 109
Photosensitizer 14, 20
Pinhole 108, 115, 117
Planar zig-zag conformation 80
Plank's constant 47
Poly(tetramethylene oxide) 43, 79-84
Polycondensation 166
Polydisperse 107, 109, 118-121, 124
Polydispersity index 107
Polyelectrolyte 18, 22, 33
Polyethylene 48-79
–, high-pressure crystallized, phase structure 64-70
–, phase structure, randomly distributed ethyl branches 74-79
–, – –, crystallized from the melt 48-61
–, – –, crystallized from dilute solution 61-64
–, ultra high modulus, phase structure 70-74
Polymer characterization 129
Polymer colloids 129
Polymerization degree 139-146, 149, 150, 157, 161, 169
Polymers, homologous 118
–, thermoplastic 128
Polypropylene gel, syndiotactic 89-98
Polystyrene 90
Polyvinyl alcohol 43, 90
Position-sensitive detector 115
Post-gelation analysis 137, 155, 162-165
Principal values 44
–, chemical shift tensor 51
Proton nutation 84
Proton spin-lattice relaxation time, T_{1H} 57
Pseudobrush 15, 16
Pulse sequence 45

Quasi-elastic light scattering (QELS) 109

α-Relaxation 49
β-Relaxation 49
γ-Relaxation 49

Radius of gyration 15, 104, 108, 136
Raman spectroscopy 42, 76, 77
Random case 158
Random evolution 154
Rate constant 138, 144-148, 156-159, 166, 170
Reaction paths 136
Reaction time 141
Reactions of growth 139
Recursive solution 160
Redox system 7
Refractive index 104-108, 118
Refractive index increment 108, 115, 124, 125
– – –, specific 108, 115
Refractometer 103, 115, 117
Regularly folding molecular chains 63
Repetition time 46
Rescaled time 139
Resonance frequency 47
Resonant oscillating field, B_1 44
Reverse transcription-polymerase chain reaction 24
Rheumatoid arthritis 29
Rheumatoid factor 29
Riccati differential equation 161
Rigid rod-like chain 118
Rigidex 64
Ring-opening polymerization 80
Rubbery state 42

Saturation-recovery pulse sequence 52
Scaling 138, 168, 169
Scattered light intensity 107, 122
Scattering, depolarized 109
Scattering angle 108, 109, 127
Scattering intensity, time-averaged 103, 104, 108, 118
Sedimentation equilibrium 107
Selection rule 93
Self-consistent field theory 15
Signal/noise ratio 45, 64, 100, 111
Single correlation-time theory 48
Size distribution 129
Size exclusion chromatography 122
Small-angle X-ray scatttering 42
Smoluchowski-like equation 158-162, 167
Sol 162-164
Solid extrusion 74
Solid-solid crystal transformation 73
Solution-grown polyethylene 62
Solution-mediated wet method 34
Spectral densitiy 46, 47
Spin relaxation 46
Spin-lattice relaxation, T_1 relaxation 46
Spin-spin relaxation, T_2 relaxation 46
sPP/carbon disulfide gel 93

Subject Index

sPP/*o*-dichlorobenzene gel 51, 90, 96
sPP gel 89-98
Stacked lamellar structure 58, 59
Static LLS 105, 107, 108, 115, 122, 124, 127, 128
Statistical analysis 137, 142-145, 152, 155, 156, 160-163
Statistically generated distribution 155
Step growth 137, 139, 145, 146, 150, 151
Steric hindrance 154
Stick-type spectrum 80, 85
Stochastic Markov process 156
Stoichiometric mixture 150
Stokes radius 129
Stokes-Einstein equation 16
Straight line decomposition 48
Substitution degree 143, 147, 156, 157, 160
Substitution effect 143, 146, 147, 151, 156-160, 167
Supercooled state 42
Symmetrical decomposition 48
Symmetry 136, 139, 140

Tenacious adhesion 32
Tensor of rank 44
Tertamethylsilane (TMS) 51
Tetrahydrofuran 80
$T^H_{1\rho}$, ^1H spin-lattice relaxation time in rotating frame 79
Thermal equilibrium spectrum 46, 75
Thermal equilibrium state 51
Threefold jump rotation 86
Threefold rotational motion 79, 86, 87
Torchia_s pulse sequence 52, 81
Trans-trans methylene sequence 51
Trans-trans-gauche-gauche molecular sequence (ttgg form) 90

Translational diffusion coefficient 103, 104, 109, 119, 122, 127
Translational diffusion coefficient distribution 110
Transverse relaxation 46
Tree 139, 160, 170
Triethyloxonium hexafluoroantimonate 80
Triplet state 12
ttgg crystal 90-92
Two-phase structure 86

UHMPE 70-74
Ultra high modulus polyethylene (UHMPE) 70-74
Ultracentrifugation 111, 129
Uniaxially drawn sample 73
Unreacted group 143, 157

van der Waals energy 18
Vapour pressure osmometry 107
Vibrational spectroscopy 90
Viscometry 107
Viscosity 137
Viscosity-average molecular weight 48

Ways of assembling 136, 155
Weight distribution, molecular weight 118
– –, resulting cumulative 122
Weight fraction 105, 118, 124, 125, 127, 128

X-ray diffraction analysis 80, 85, 86

Z-average molecular weight 107
Zeeman energy, H_Z 43, 44
Zero frequency component 48
Zimm plot 108

Springer and the environment

At Springer we firmly believe that an international science publisher has a special obligation to the environment, and our corporate policies consistently reflect this conviction.

We also expect our business partners – paper mills, printers, packaging manufacturers, etc. – to commit themselves to using materials and production processes that do not harm the environment. The paper in this book is made from low- or no-chlorine pulp and is acid free, in conformance with international standards for paper permanency.

Printing: Saladruck, Berlin
Binding: Buchbinderei Lüderitz & Bauer, Berlin

T
1 Month
green dot